AF763515

Materials Forming, Machining and Tribology

Series editor

J. Paulo Davim, Aveiro, Portugal

More information about this series at http://www.springer.com/series/11181

Kapil Gupta · Neelesh Kumar Jain

Near-Net Shape Manufacturing of Miniature Spur Gears by Wire Spark Erosion Machining

Springer

Kapil Gupta
Department of Mechanical and Industrial Engineering Technology
University of Johannesburg, Kingsway Campus
Johannesburg
South Africa

Neelesh Kumar Jain
Discipline of Mechanical Engineering
Indian Institute of Technology Indore
Indore
India

ISSN 2195-0911 ISSN 2195-092X (electronic)
Materials Forming, Machining and Tribology
ISBN 978-981-10-1562-5 ISBN 978-981-10-1563-2 (eBook)
DOI 10.1007/978-981-10-1563-2

Library of Congress Control Number: 2016942498

© Springer Science+Business Media Singapore 2017
This work is subject to copyright. All rights are reserved by the Publisher, whether the whole or part of the material is concerned, specifically the rights of translation, reprinting, reuse of illustrations, recitation, broadcasting, reproduction on microfilms or in any other physical way, and transmission or information storage and retrieval, electronic adaptation, computer software, or by similar or dissimilar methodology now known or hereafter developed.
The use of general descriptive names, registered names, trademarks, service marks, etc. in this publication does not imply, even in the absence of a specific statement, that such names are exempt from the relevant protective laws and regulations and therefore free for general use.
The publisher, the authors and the editors are safe to assume that the advice and information in this book are believed to be true and accurate at the date of publication. Neither the publisher nor the authors or the editors give a warranty, express or implied, with respect to the material contained herein or for any errors or omissions that may have been made.

Printed on acid-free paper

This Springer imprint is published by Springer Nature
The registered company is Springer Science+Business Media Singapore Pte Ltd.

Preface

In recent years, the demand of micro-components and tiny parts has increased rapidly due to the growing trend of miniaturization. Technological advances have aided the production of miniaturized devices by making it possible to create high-quality micro-parts. Miniature gears are an important and an extensively used micro-component with widespread applications. These applications range from MEMS-NEMS, miniature pumps, miniature motors, timer mechanisms to robots, home appliances, electronics, cars, and even aircrafts.

The main purpose of the use of miniature gears was the transmission of motion and/or power. The functional performance, reliability, and service life of machines and devices depend largely on the manufacturing quality, i.e., dimension, form, and surface quality of the miniature gears employed. The size of these gears ranges from 0.1 to 10 mm in outside diameter. High manufacturing quality of gears ensures smooth, accurate, and noise-less operation. This necessitates highly precise and accurate processes to manufacture these gears. Conventional processes (i.e., hobbing, stamping, die-casting, powder-metallurgy, and extrusion) are productive and give good surface finish, but they yield poor quality (DIN standard 9–12) in terms of micro-geometry/form features. Consequently, secondary finishing and sizing operations are required to manufacture net-shaped miniature gears. The inclusion of secondary operations heavily increases the cost of gear manufacture, lengthens the production chain, and consequently causes high environmental footprint.

Recently, spark-erosion-based processes such as spark erosion machining (SEM)\wire-SEM (WSEM) and their micro-versions have been explored to overcome the limitations of the conventional processes for miniature gear manufacturing. The research work detailed in this book aims to explore and establish WSEM as a near-net-shape process to manufacture high-quality fine-pitched miniature spur gears with improved dimensional accuracy, micro-geometry, and surface quality. This involved properly planned detailed experimental investigations in four stages, careful analysis and interpretation of the results, development of models for micro-geometry, surface roughness and gear cutting rate using regression analysis and artificial neural network (ANN), and multi-objective optimization of the WSEM process parameters (for high-quality miniature gears and productivity of WSEM process) using response surface modeling-based desirability analysis and back propagation neural network integrated genetic algorithm techniques.

Significant achievements of this research work are: (i) attainment of gear quality up to DIN standard 5 which is much superior than produced by any conventional manufacturing process for the miniature gears; (ii) average roughness up to 1.05 μm and maximum roughness up to 6.34 μm with good running-in properties; (iii) good surface integrity aspects having crack-free microstructure and very thin recast layer; (iv) establishing superiority of WSEM over hobbing for manufacturing of high-quality miniature gears; and (v) establishing WSEM as a near-net-shape manufacturing alternative to the conventional processes for miniature gears.

This will significantly improve the operating performance (noise-less and accurate transmission) and enhance the service life (better tribology aspects and wear characteristics) of the miniature gears.

This work certainly opens up many avenues for future research on miniature gear manufacturing by wire spark erosion machining (WSEM). Some worth-mentioning are: exploring WSEM for manufacturing high-quality miniature helical, bevel, internal, and planetary gears; investigating other commonly used miniature gear materials such as stainless steel, bronze, aluminum; investigating the effects of WSEM process parameters on noise, vibration, and wear rate of the miniature gears; employing WSEM for gear tooth modifications such as tip correction, root alteration, and crowning.

This book is a good source of information on miniature gears, wire spark erosion machining, and its capability to manufacture net-shaped or near-net-shaped miniature gears. It enables the researchers and engineers by providing them a vast knowledge of the above-mentioned fields, which certainly helps them to explore future avenues in manufacturing high-quality miniature gears for wide application requirements.

It opens with Chap. 1 introduces miniature gears and discusses various parameters to determine their quality and service life, and conventional processes to manufacture them. Chapter 2 presents an insight into wire spark erosion machining process and a detailed review of past work done on manufacturing miniature gears using spark erosion and wire spark erosion machining processes. It also mentions the research gaps evident from this review and furthermore describes the identified research objectives of the present work. Chapter 3 describes the planning, design, details, and procedure of experimental investigations for different stages, i.e., trial, pilot, main, and confirmation experiments. It also describes evaluation procedure of the measures of gear quality and WSEM productivity. This chapter also mentions the methods and tools employed for analysis of the experimental data. Chapter 4 presents the experimental results, their analyses, and conclusions for each experimental stage. Chapter 5 describes the modeling by regression analysis and artificial neural network, and optimization of WSEM process by desirability analysis and genetic algorithm. A comparative study between WSEM and hobbing to manufacture miniature gear of the same specifications is given in Chap. 6, which also highlights the significant achievements and conclusions from the present research work.

Finally, the chapter and conclusively the book end up with stating the avenues for future research.

We hope the book could disseminate the knowledge of wire spark erosion machining of gears and fulfill the requirements of the researchers, engineers, and academicians indulged therein.

Kapil Gupta
Neelesh Kumar Jain

Contents

About the Authors

Kapil Gupta is a Postdoctoral Research Fellow in the Department of Mechanical Engineering Science at University of Johannesburg. He obtained his Ph.D. in mechanical engineering with specialization in Manufacturing Engineering from Indian Institute of Technology Indore, India, in 2014. Advanced machining processes, sustainable manufacturing, precision engineering, and gear technology are his areas of interests. Kapil Gupta has authored eight SCI Journal articles based on advanced and sustainable manufacturing, one book on hybrid machining processes, and also delivered presentations in some prestigious international conferences. He possesses five years of professional experience that includes teaching and research.

Neelesh Kumar Jain is currently a Professor of Mechanical Engineering and Dean Academic Affairs at Indian Institute of Technology Indore, India. He obtained his Ph.D. in Mechanical Engineering from Indian Institute of Technology Kanpur, India, in 2003. His areas of research interest include advanced and hybrid machining processes and gear technology. Dr. Jain has successfully supervised several Ph.D. scholars and masters and undergraduate candidates. He is the author of over hundred published scholarly articles. He is also serving as the Head of Centre of Excellence (an international research lab) in Gear Engineering at IIT Indore.

Chapter 1
Introduction

1.1 Introduction to Miniature Gears

Miniaturization, one of the key requirements of the present technological era is significantly influencing engineers and technologists globally to explore new techniques i.e. advanced, hybrid and sustainable processes for manufacturing miniaturized products and their components. *Miniature gear* is one of the most important miniaturized components and is extensively utilized in micro-electro-mechanical-systems (MEMS), scientific instruments, miniature motors, robots, miniature pumps, timer mechanisms, automobiles, medical-equipment, toys, business machines, portable machine tools, and various other electronic-appliances (Fig. 1.1 shows some of the important application domains of miniature gears). The functional characteristics of these devices largely depend on the manufacturing quality of miniature gears used. Therefore, highly precise and accurate processes i.e. *near-net shape* manufacturing processes are required to manufacture quality gears. The term *near-net shape manufacturing* implies that the initial production of the engineered part or component is very close to the final (net) shape, reducing the need for subsequent treatment i.e. finishing etc. This saves a significant amount of production costs and simultaneously keeps the environment green and clean.

Selection of a suitable gear material is also important to achieve the desired performance of miniature gears. Commonly used miniature gear material can either be ferrous (stainless-steel), non-ferrous (brass, bronze, aluminum) or plastics [1, 2]. Ferrous materials offer high strength; therefore, they are generally used as load carrying or torque transmission gears. Non-ferrous materials offer good machinability, light weight and corrosion resistance. Motion-carrying gears are generally made of non-ferrous materials or plastics. Fundamentally, poor machining characteristics and temperature instability limits the applicability of plastics.

Minimum running noise, accurate motion transfer, sufficient torque transfer capability and long service life are the key functional requirements for miniature gears. In order to achieve these characteristics, the gear should have good surface

© Springer Science+Business Media Singapore 2017

K. Gupta and N.K. Jain, *Near-Net Shape Manufacturing of Miniature Spur Gears by Wire Spark Erosion Machining*, Materials Forming, Machining and Tribology, DOI 10.1007/978-981-10-1563-2_1

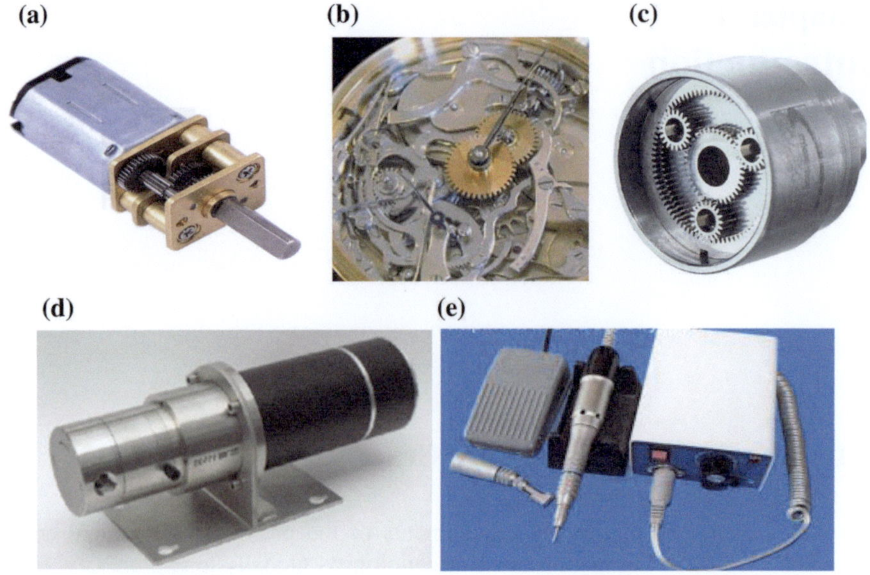

Fig. 1.1 Important applications of miniature gears in **a** Micro-motors; **b** Sophisticated watches and timer mechanisms; **c** Micro planetary gearboxes used in micro-reducers and harmonic drives; **d** Micro-pumps; and **e** Bio-medical instruments (the figure displayed is of an instrument used for Dental Surgery)

finish (teeth surface should have minimum roughness and free from nicks, dirt, burrs and other defects), be geometrically accurate (have low geometric tolerance) and superior near surface mechanical integrity (high hardness). These requirements may be achieved by appropriate selection of the material, heat treatment conditions, and manufacturing and/or finishing processes.

Miniaturization is a relative concept that implies "making something smaller". The actual concept of what constitutes small or miniature must therefore be defined unambiguously. A miniature engineering system also called microsystem usually contains MEMS components that perform specific engineering functions. A MEMS may contain components of sizes ranging from 1 to 1000 μm (1 mm) with the microsystem itself being of the mesoscale. Mesoscale implies a scale that is between micro and macro scales i.e. it is the size range of millimeters to a centimeter [3]. In a keynote paper during a meeting of the Scientific Technical Committee of the Physical and Chemical Machining Processes of *CIRP* in 2000, *Masuzawa* [4] suggested that the 'micro' in micromachining implies sizes between 1 and 999 μm. In his specific keynote a range of 1–500 μm was adopted to refer to micro. He noted that the term *micro* may vary according to era, person, machining technique, type of product and material. This implies that currently there is no consistent international classification on the basis of size or dimensions for miniature components including gears. To be consistent miniature gears are therefore categorized as either micro-gears (outside diameter less than 1 mm) or meso gears (outside diameter 1–10 mm) [3, 5, 6].

This book discusses the aspects of spark erosion machining of meso-sized gears made of brass. These gears are fine-pitched gears generally employed to run at very high speed (up to 50,000 rpm) and are primarily used for motion transmission. Consequently, minimum running noise, accurate motion transfer, and long service life are mainly the important desirable characteristics for these gears.

1.2 Quality Aspects of Miniature Gears

The functional performance and service life of a gear mainly depend on its material and post manufacture surface integrity. Therefore, after careful selection of material and method of manufacture, gears should have better surface integrity aspects for optimum operating performance during their service life. In general, surface integrity parameters can be divided into two categories (i) external features of the surface such as surface roughness, waviness, form errors, surface topography etc.; and (ii) metallurgical and mechanical features of the internal sub-surfaces such as microstructure, micro-hardness, and residual stresses etc. [6–12].

The most significant surface condition indicators for gears are; form and location errors, flank surface topography, surface texture or roughness parameters, and micro-hardness and microstructure of the gear tooth flanks. The important surface roughness parameters which govern the tribological behavior and largely determine the service life of a gear include average surface roughness, maximum surface roughness, bearing area properties, skewness and kurtosis [8–12]. Inaccuracy in the micro-geometry and macro geometry of a gear causes deviation from the ideal motion transmission conditions. The errors or deviations in micro-geometry i.e. form and location errors greatly affect the operating performance of gears; these include errors or deviations in the profile, lead, pitch and runout [1, 7, 8]. The profile error affects the noise behavior, the lead error governs the load carrying capacity, whereas pitch error and runout affect the motion transfer characteristics as depicted in the Fig. 1.2. Chordal tooth thickness, span and outside diameter of a gear are important macro-geometry parameters that also affect its functional performance to some extent.

Two important international standards define the manufacturing quality of gears based on the tolerance specified for micro-geometry errors are: (i) Deutsche Normen (*DIN*), the German standard [13], and (ii) American Gear Manufacturers Association (*AGMA*), the American standard. The DIN standard has range of 1–12 while, AGMA standard ranges from 1–15. Lower DIN number or higher AGMA number indicates better quality of the gear and vice versa. Table 1.1 presents the quality requirements of the gears for various applications in terms of DIN and AGMA numbers along with the manufacturing processes used. The DIN standard is referred to define the quality of miniature gears fabricated during the research work discussed in this book.

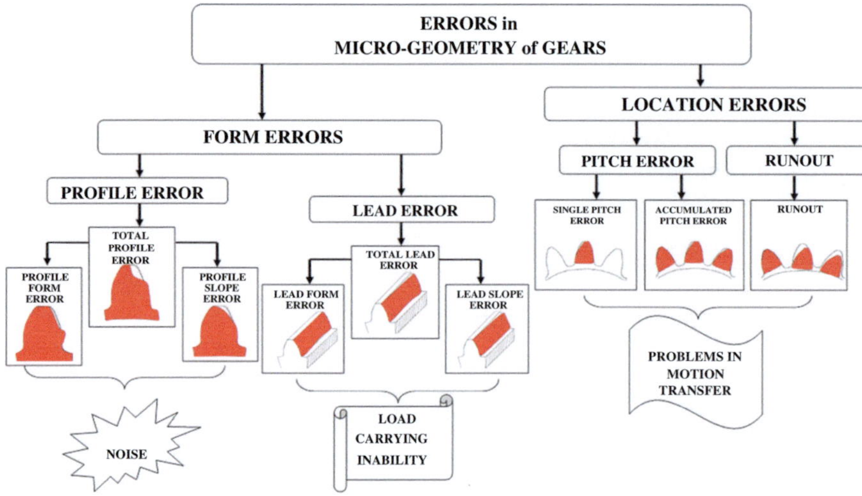

Fig. 1.2 Details of various form and location errors of a gear [8]

Table 1.1 Quality requirements for various gear applications in terms of DIN and AGMA standards [1, 2, 14]

Application type	Typical examples	AGMA quality number	DIN quality number	Corresponding manufacturing or finishing process
Commercial applications	Hand tools, pumps, clocks, slow speed machineries, various appliances	3		Plaster-mold casting, permanent-mold casting
		4	12	Investment casting, *injection molding**, *extrusion**
		5	11	*Die casting**
		6	10	Milling, *stamping**, *powder metallurgy**
		7		Rolling, broaching
Precision applications	Aircraft engines, turbines, cameras, automatic transmission systems, instruments, high speed machineries	8	9–10	Rolling, shaping, *hobbing**
		9	8–9	Rolling, shaving, honing, lapping, grinding
		10	7–8	Shaving, honing, lapping, grinding
		11	6–7	Shaving, grinding
		12–13	4–6	Grinding
Ultra-precision applications	Precision instruments, military navigations	14	3–4	Grinding

*Used for manufacturing of metallic *meso-gears*

1.2.1 Micro-geometry

The important form parameters, i.e., profile and lead; and the parameters define the angular (pitch) and radial locations of teeth (i.e. runout) constitute gear's micro-geometry. The errors or deviations in these parameters are mainly termed as *form errors* and *location errors*. Figure 1.2 depicts these two classes and corresponding types of micro-geometry errors for a gear.

1.2.1.1 Form Errors

Form errors are the first and second order deviations from the intended nominal shape of a surface as depicted in Fig. 1.3 [15]. These are broadly spaced surface irregularities of low frequency and long wavelength. As mentioned, the form errors of a gear include profile error and lead error.

Profile error is the difference between the measured and the theoretical involute surfaces (as shown in Fig. 1.4a) and it determines the noise characteristics of the gears. It has two components namely profile form error and profile angle (slope) error which are measured perpendicular to the functional profile. Functional profile is that portion of the tooth profile which remains in contact with the meshing tooth.

Profile form error (ff_a) is the difference between the nominal involute form and the actual involute form, while *profile angle error* (fh_a) is the difference between

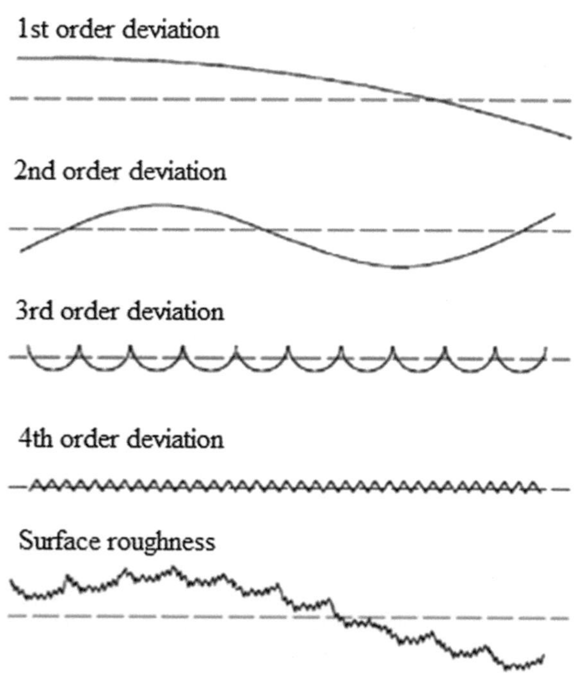

Fig. 1.3 Deviations from the nominal shape of a surface (DIN 4760 Standard 1982 [15])

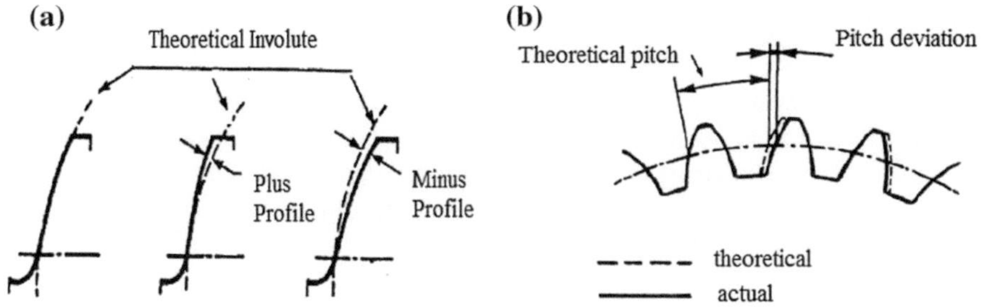

Fig. 1.4 Concept of errors in **a** profile; and **b** pitch, for an external gear [6], with kind permission from Elsevier

the nominal involute angle and the actual involute angle. The *total profile error* (F_a) is the sum of profile form and profile angle errors.

Lead error is the deviation of the actual surface from the theoretical surface of a gear tooth along the face width and it governs the load carrying capacity of the gears. It has two components namely *lead form error* (ff_β) and *lead angle* (*slope*) *error* (fh_β) which are measured at the middle of the tooth height along the face width of the gear. *Lead form error* is the difference between the nominal lead form line and the actual form line, whereas *lead angle error* is the difference between the nominal helix angle and the actual helix angle. The *total lead error* (F_β) is the sum of lead form and lead angle errors.

1.2.1.2 Location Errors

Location errors define the accuracy of location or position of teeth on a gear. They are of two types namely pitch error and runout. Pitch is the distance between the two corresponding points on two consecutive gear flanks on the same side along the pitch circle. *Pitch error* describes the inaccuracy in angular location of the gear teeth along the pitch circle of the gear. Pitch error is of two types i.e. single pitch error (f_p) and accumulated or total pitch error (F_p), which are measured at the middle of the tooth height along the pitch circle. *Single pitch error* is the algebraic difference between the theoretical and actually measured values of the pitch for a pair of teeth as depicted in Fig. 1.4b. *Accumulated or total pitch error* is the maximum value of location inaccuracy between any two teeth of a gear along its pitch circle. It is also known as *index error.*

Runout describes the inaccuracy in the radial location of the gear teeth with reference to the pitch circle. It is the maximum difference between the nominal or theoretical radial position of all teeth to the actual measured position and is also measured at the middle of the tooth height. The form and location errors are the measure of geometric inaccuracy of a gear and based on the amount of these errors, the gears are categorized in and assigned certain quality numbers.

Miniature brass gears are primarily used for motion transfer and subjected to very high speed and low torque conditions; therefore, being the motion-transmission gears, noiseless operation and accuracy in transmission are the two major requirements for these gears. Consequently, profile and pitch are the more important micro-geometry parameters which need special attention for these types of gears.

1.2.2 Macro-geometry

Macro-geometry parameters namely span, chordal tooth thickness, and outside diameter govern the functional characteristics of gears which in turn affect their operating performance [1, 16]. *Span* is the distance across certain number of teeth along a line tangent to the base circle. *Span deviation* is the difference between its theoretical value and actually measured value. *Chordal tooth thickness* is given by 0.5 ($\pi \times$ *module*). The deviation in chordal tooth thickness is the difference between its theoretical and actual values. *Outside diameter* is the maximum distance between the diametrically opposed tooth spaces of a gear and measured when two balls of the same size are being accommodated in these tooth spaces. Deviation in outside diameter is the difference between its theoretical and actual values.

1.2.3 Tooth Flank Topography

Investigations on the *gear tooth flank surface topography* reveals the presence of nicks, burrs, asperities or peaks and valleys and it is important to predict the tribological behavior, running-in period and service life of a gear. The topography is represented by two slanted perspective planes which give three-dimensional impression of the gear tooth flank surface. The image consists of several profile and/or lead measurements that are uniformly distributed over the face width of the gear tooth such that an apparently three-dimensional surface pattern is obtained as shown in Fig. 1.5. The measured points are joined with straight lines in the profile and/or lead directions. The number of points used depends on the desired resolution.

1.2.4 Surface Roughness

Surface finish is a colloquial term widely used to denote the general quality of a surface. It is not specifically tied to the texture or specific surface roughness values. However, a good surface finish implies low surface roughness and vice versa. Surface roughness refers to the higher order surface irregularities of short-wavelength and high frequency as depicted in Fig. 1.3 [15]. It is mainly

Fig. 1.5 Representation of gear tooth flank topography [12]

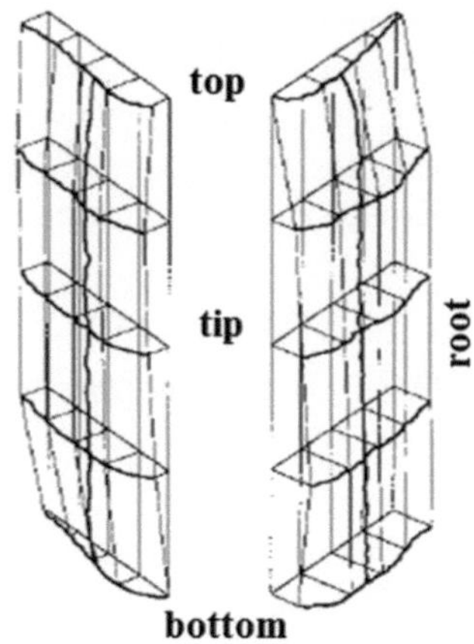

concerned with random asperities, scratches and tears that occur at small wavelength and caused by the nature and actions of the manufacturing processes [17]. Surface finish of a gear governs its wear characteristics and service life [10, 11]. Surface roughness parameters are normally categorized into four groups, i.e. amplitude parameters, statistical parameters, spacing parameters and hybrid parameters. The present work is concerned with amplitude and statistical parameters of the miniature gear tooth surface. Average surface roughness, maximum surface roughness (maximum peak-to-valley height) and mean roughness depth are the important amplitude parameters.

Average surface roughness (R_a) is the arithmetic average of the absolute values of roughness profile ordinates (Fig. 1.6). It is computed using Eq. 1.1.

$$R_a = \frac{1}{l}\int_0^1 |Z(x)|dx \tag{1.1}$$

Maximum surface roughness (R_{max}) is the distance between the highest peak and deepest valley within an evaluation length (Fig. 1.6).

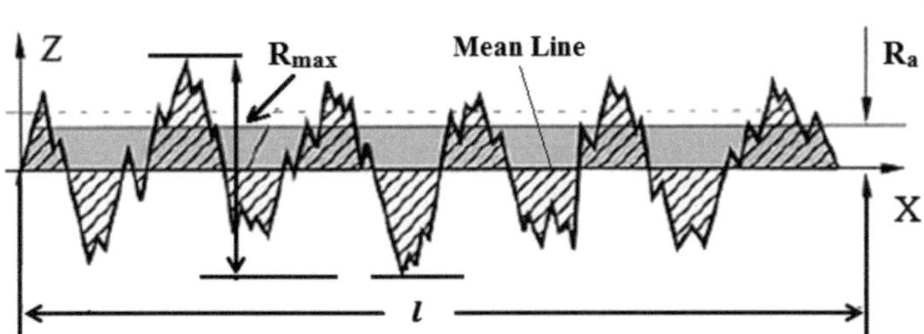

Fig. 1.6 Representation of roughness parameters: average roughness 'R_a' and maximum roughness 'R_{max}'

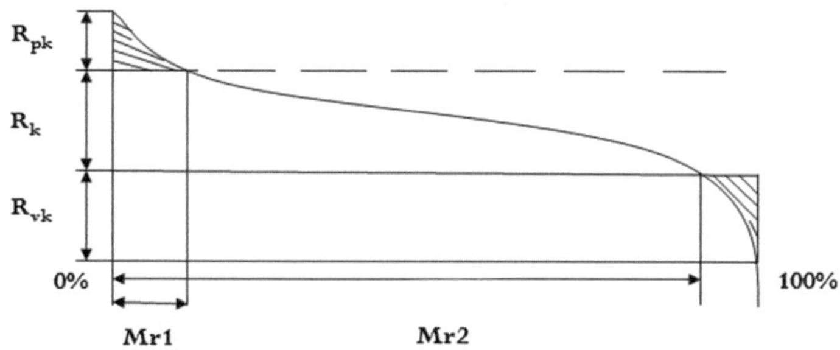

Fig. 1.7 Concept of Abbott-Firestone or bearing area curve [12]

An important statistical parameter of the surface roughness which describes tribological or bearing properties of a functional surface is the *Abbott-Firestone curve* or *material ratio curve* or *bearing area curve* (*BAC*). It is a graph between maximum surface roughness R_{max} and the percentage evaluation length. Figure 1.7 depicts the concepts underlying in a bearing area curve with its relevant parameters. It describes the contact area with a moving imaginary plane as a function of the distance i.e. at certain depth from the surface peaks. This curve describes the amount of material which would be in contact as the percentage of the evaluation length. Flatter the BAC better is the uniformity of the surface finish.

Referring to Fig. 1.7, *reduced peak height* 'R_{pk}' is the height of the top portion of the surface roughness profile and the material component corresponding to the upper limit of the roughness core 'M_{r1}' indicates the area of the initial surface roughness profile which is likely to wear out during initial run-in period. *Core roughness depth* 'R_k' is the depth of the surface roughness profile excluding prominent peaks and valleys. After initial run-in period R_k influences the life and performance of the gears. Material component corresponding to the lower limit of the roughness core is 'M_{r2}'. *Reduced valley depth* 'R_{vk}' is depth of the lowest part of the surface roughness profile which shows the lubricating oil retaining capacity.

Skewness and *kurtosis* are the other two important statistical parameters of surface roughness that govern the tribological characteristics of the contacting surfaces [9]. *Skewness* 'S_{sk}' is a measure of symmetry of the variations of a surface about the mean line (Fig. 1.8a). A surface with positive value of skewness has a wider range of peak heights. Negative skewness is the criterion for a good bearing surface. *Kurtosis* 'S_{ku}' defines sharpness of the peaks of asperities or the surface height distribution (Fig. 1.8b). Low kurtosis value (<3) corresponds to broad tip, whereas the opposite (>3) characterizes the sharp hills.

High surface roughness of the gear teeth leads to early failure due to wear in the form of micro and macro pitting, spalling, and scoring. Therefore, surface roughness should be minimized to prevent premature failures of the miniature gears.

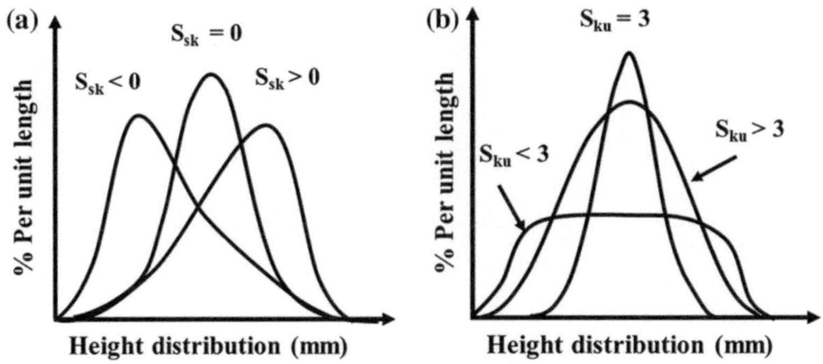

Fig. 1.8 Representation of **a** skewness and **b** kurtosis parameters [8]

1.2.5 Microstructure and Micro-hardness

The presence of nicks, burrs, micro-cracks, voids, and recast layer; and the effects of metallurgical, thermal, mechanical and chemical changes on any machined surface can be determined by examining its microstructure through optical microscopy or scanned electron microscopy or atomic force microscopy. The top layer of the parts machined by spark erosion machining or wires spark erosion machining may have a white layer known as the *recast layer* formed by the solidification of un-expelled molten material [18–21]. It has different microstructure from its parent material and consists of pockmarks and voids on its surface. Micro-cracks can form easily on this very hard and brittle recast layer. Heat affected zone (HAZ) exists immediately after the recast layer which is due to rapid heating-quenching cycles. Figure 1.9 depicts alteration of the metal surface due to the effect of spark erosion machining.

The thickness of recast layer and the depth of HAZ depend on the heat sinking ability of the work material and amount of the discharge energy used. Sometimes,

Fig. 1.9 Recast layer and heat affected zone (HAZ) for a spark erosion machined surface

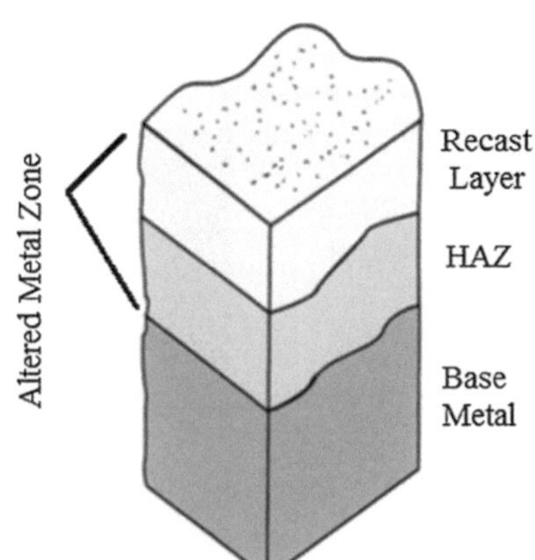

it is difficult to know the presence and thickness of recast layer through microstructure examination; therefore, micro-hardness testing can be used to identify them. The recast layer increases the hardness of the upper surface, while the hardness of the core material remains unaltered and this can be easily examined by micro-hardness testing of spark eroded parts.

1.3 Conventional Methods of Miniature Gear Manufacture

Manufacturing techniques such as lithography (also named as LIGA), micro-hot embossing and micro-electro forming etc. are typically used to manufacture micro-gears only [3, 5]. Moreover, these methods are complex, expensive and not suitable for all types of gear shapes and materials [8, 22, 23]. Since, the present research work is concerned with the meso-gears made of brass; therefore, all the conventional processes used to manufacture meso-gears were studied so as to understand their limitations and capabilities, and to compare them with the wire spark erosion machining process. Processes for manufacturing the metallic meso-gears can be broadly categorized into three groups, namely, (i) Material removal processes: hobbing; (ii) Forming processes: stamping and extrusion; and (iii) Additive processes: die casting and powder metallurgy. All these conventional processes also suffer from some inherent limitations individually and in common, all these processes produce poor quality gears that usually exposed to subsequent finishing operations such as grinding, lapping, and honing etc. [1, 2, 6, 8]. In other words, none of the conventional process alone is capable to manufacture net shaped or near-net shaped gears. Table 1.2 summarizes the major limitations of these conventional processes along with the quality of miniature gears produced by them.

Following sections briefly describe the working principle of these processes along with material and product specific applications of miniature gears manufactured by them.

1.3.1 Material Removal Processes

Hobbing is the most commonly used material removal process for manufacturing miniature gears. It is a generating process in which both, the cutting tool called 'hob' and the gear blank rotate in a constant relationship with simultaneous feeding of the hob into the gear blank. As the hob is fed across the blank, all the teeth in the gear are completely cut. In other words, it is a machining process in which gear teeth are progressively generated by a series of cuts with a hob [1, 2, 24]. This process is capable to manufacture gears, splines, ratchets and sprockets. It is used to manufacture spur, helical and worm gears but cannot manufacture bevel or internal

Table 1.2 Limitations of the conventional processes used to manufacture miniature gears [1, 2, 6]

Manufacturing process	Limitations	Gear quality (DIN No.)
Hobbing	• Generates tool marks on gear teeth • Requires subsequent finishing operation for further improvement in gear quality • Long setup time	9
Stamping	• Shaving operation is needed for final finishing • Gears made by this process have tooth thickness limitations • Wear and tear of die-punch is a major problem • Applicable for no load to medium duty gears only	10
Extrusion	• To get the accuracy required for gears, a secondary drawing operation is required after extrusion • Wear of die is problematic • Not suitable for fine pitch gears • Limited to spur gears only	12
Die casting	• Requires repair and maintenance of die at frequent intervals • Cannot be used for extremely accurate gears • Trimming operations are necessary after the gear has been removed from the die	11
Powder metallurgy	• Arrangement of fine metal powder of all type is difficult • Gears are subjected to repressing or coining after sintering to correct the edges • Secondary operations such as shaving, burnishing, and grinding are required to obtain high quality	10

gears. Hobbing can be used for any gear material and is the most economical for medium and high volume production of gears. But, the quality of hobbed gears is poor (up to DIN 9) thus necessitating secondary finishing operations such as grinding, shaving and honing. Hobbed miniature gears are mainly used in the miniature motors and pumps, machine tools, timing devices, and automotive parts etc.

1.3.2 Forming Processes

Stamping and *extrusion* are the most commonly used forming processes for miniature gear manufacturing. In *stamping* process, a sheet of metal is placed between the top and bottom portion of a die, the upper die is pressed into the lower section and removes or cuts the gear from the sheet [1, 2]. This is a low-cost and an efficient method for producing lightweight gears for no-load to medium-duty applications. Stamping is restricted to the thickness of the gear in the range from 0.25 to 2.5 mm only and is primarily used to manufacture spur gears. This method is economical only for high production volume. When greater face widths are required, a number of individual blanked pieces are laminated together and fastened by riveting, press fitting, or welding. The shearing action between die and punch in stamping causes the teeth and outside periphery of gear to be somewhat rough and rounded at one edge. This rounding is called drawdown which may affect the performance of gears adversely. The rate of stamping ranges from 35 to 200 pieces per minute. Additional shaving operation is required to get stamped gears with less drawdown and roughness. A wide range of materials can be processed by stamping, including all the low and medium carbon steels, brasses, and some aluminum alloys. Stamped miniature gears are used in electric and water meters, toys, clocks and timer mechanisms, watches, mixers, blenders, toasters, washers and dryers etc.

In *extrusion* of gears, a bar is pushed or extruded through a series of several dies in which the last die has the final shape of the desired tooth form. As the material passes through these dies, it is actually squeezed into the shape of the die [1, 2]. Since, the material is displaced by pressure; the outside surface is work hardened and quite smooth. After passing through the dies, the formed gears are known as pinion rods, and often are put into screw machines that finish the individual gears. Any material that has good drawing properties, such as high-carbon steels, brass, bronze, aluminum, and stainless steel, may be used for the drawn pinion rod. Extruded miniature gears have a large variety of applications such as watches, clocks, typewriters, carburetors, magnetos, small motors, switch apparatus, taximeters, cameras, slot machines, all types of mechanical toys, and many other parts for machinery of all kinds. To get the accuracy required for gears, a secondary drawing operation is required after extrusion. The process is limited to straight spur gears and is economical when quantities are moderate or greater.

1.3.3 Additive Processes

Die casting and *powder metallurgy* are the most commonly used additive processes for miniature gear manufacturing. Theoretically, any casting process can be used to produce any gear to semi-finished state, but only die casting process gives acceptable dimensional accuracy and surface finish. Casting of gear blanks made of steel, bronze, and other materials is quite common. *Die casting* involves forcing the molten metal under pressure into the die cavity. The material is then allowed to solidify and removed from the die [1, 2]. Several materials such as zinc, brass, aluminum and magnesium can be used to manufacture gears by the die cast process. This process is suitable to high production volume of low-cost gears. Spur, helical, worm, bevel, and shouldered and stepped gears can be die cast. It is widely used for the production of miniature gears used in the home appliances, business machines, hand tools, instruments, cameras, etc. Trimming operations such as shaving or broaching are necessary after removing a gear from the die.

For manufacturing of gears by *powder metallurgy* process, a homogeneous mixture of materials in the powdered form is compacted in the die and then sintered at a temperature below the melting point temperature of the constituent materials under a controlled atmosphere in a furnace that binds the materials together [1, 2]. The gear shape coming out from the furnace is subjected to coining operation. A wide variety of base materials are available in powder form, such as brass, bronze, iron, low alloy steels, and stainless steel. Powder metallurgy gears are used in the home appliances, office machines, machine tools, and automotive products. Gears other than spur type are difficult to manufacture by this process. This process alone cannot make quality gears and needs subsequent finishing operations.

Conclusively, based on the limitations of versatility and capability of the conventional processes, and the quality of gears produced by them, it can be stated that all the conventional processes of miniature gear manufacturing produce gears of low quality having DIN quality number in the range of 9–12, thus necessitate the assistance of some subsequent finishing operation(s). These subsequent finishing operations themselves require fabrication, repair and maintenance of the finishing tool; consumes large amounts of cutting fluid and energy; increases the burden of handling, recycling and disposal of waste and escalates the overall cost. This has led to numerous scientists and engineers exploring alternative methods to overcome the limitations of the conventional processes by producing high quality miniature gears in a single operation (i.e. through manufacturing only).

Wire spark erosion machining (WSEM) has the potential to be such process. It has always been a preferred choice for manufacture of miniaturized products, because it is capable to cut any electrically conductive material irrespective of its mechanical properties in general, and hardness and toughness in particular, gives high quality surface finish and better dimensional accuracy, generates burr-free, corrosion and wear resistant surfaces which generally does not exposed to any post-finishing operation [22–26].

The main motivation behind this research is to shorten the long process chain of gear manufacture by application of WSEM process, and producing tolerances close to net shaped or near-net shaped gears. The work discussed throughout this monograph aims to explore and establish WSEM process for manufacturing high quality miniature brass gears through properly planned and designed experimental investigations, modeling and optimization stages in pursuit of an alternative to conventional processes of miniature gear manufacturing; which in-turn ensures energy, resource and economic efficiency with less environmental footprint.

References

1. Townsend DP (2011) Dudley's gear handbook. Tata McGraw-Hill Publishing Company, New Delhi
2. Davis JR (2005) Gear materials, properties and manufacture. ASM International, Ohio
3. Hsu TR (2008) MEMS and microsystems design, manufacture and nano scale engineering. John Wiley & Sons, New Jersey
4. Masuzawa T (2000) State of the art of micromachining. Keynote Pap Ann CIRP 49(2): 473–488

5. Grzesik W (2008) Advanced machining processes of metallic materials. Elsevier, Oxford
6. Gupta K, Jain NK (2014) Analysis and optimization of micro-geometry of miniature gears manufactured by wire electric discharge machining. Precis Eng 38(4):728–737
7. Goch G (2003) Gear metrology. Ann CIRP 52(2):659–695
8. Gupta K, Jain NK, Laubscher RF (2015) Spark-erosion machining of miniature gears: a critical review. Int J Adv Manuf Technol 80(9–12):1863–1877
9. Bhushan B (2013) Introduction to tribology. Willey & Sons, New York
10. Walraven J (2003) Failure mechanisms in MEMS. Proceedings of international test conference, Charlotte
11. Zhang W, Meng G, Li H (2005) Electrostatic micro-motor and its reliability. Micro Rel 45:1230–1242
12. Gupta K, Jain NK (2014) On surface integrity of miniature spur gears manufactured by wire electrical discharge machining. Int J Adv Manuf Technol 72(9–12):1735–1745
13. DIN 3962 (1978) Standard. Tolerances for cylindrical gear teeth, Deutsche Normen, © Beuth Veriag GmbH, Berlin, Germany
14. Bralla J (1998) Design for manufacturing handbook. McGraw-Hill Publishing Company, New York
15. DIN 4760 (1982) Standard. Form deviations: Concepts; classification system. Deutsche Normen, Berlin, Germany
16. Gupta K, Jain NK (2013) Deviations in geometry of miniature gears fabricated by wire electrical discharge machining. Proceedings of international mechanical engineering congress and exposition (IMECE 2013) of ASME, V010T11A047, Nov 13–21, 2013, San Diego
17. Davim PJ (2010) Surface integrity in machining. Springer, London
18. Jain VK (2002) Advanced machining processes. Allied Publishers Pvt. Ltd., New Delhi
19. Puri AB, Bhattacharyya B (2005) Modeling and analysis of white layer depth in a wire-cut EDM process through response surface methodology. Int J Adv Manuf Technol 25:301–307
20. Ho KH, Newman ST, Rahimifard S, Allen RD (2004) State-of-the-art in wire electrical discharge machining. Int J Mach Tool Manuf 44:1247–1259
21. Benedict GF (1987) Nontraditional manufacturing processes. Marcel Dekker Inc., New York
22. Di S, Haung R, Chi G (2006) Study on micro-machining by micro-WEDM. Proceedings of the 1st international IEEE conference on nano/micro engineered and molecular systems, Zhuhai
23. Ali MY, Mohammad AS (2008) Experimental study of conventional WEDM for micro-fabrication. Mater Manuf Process 23(7):641–645
24. Gupta K, Jain NK (2014) Comparative study of wire-EDM and hobbing for manufacturing high quality miniature gears. Mater Manuf Process 29:1470–1476
25. McGeough JA (2002) Micro-machining of engineering materials. Marcel Dekker Inc., New York
26. Qin Y (2010) Micro-manufacturing engineering and technology. Elsevier, Oxford

Chapter 2
Overview of Wire Spark Erosion Machining (WSEM)

Advanced machining processes (AMPs) are well established in modern manufacturing industries as they are capable of machining most electrically conductive materials irrespective of their hardness and toughness producing complex geometries, shapes and features [1–3]. *Spark-Erosion Machining* (*SEM*) also named as *Electric-Discharge Machining* (*EDM*) is the most widely used thermal type advanced machining process. It is a controlled spark erosion process in which the mechanism of material removal is melting and vaporization by a series of repeated electrical discharges occurring between the tool electrode and the workpiece in the presence of a suitable dielectric fluid [1–3].

Wire Spark-Erosion Machining (*WSEM*) also called *Wire Electric-Discharge Machining* (*WEDM*) is a derived process of *SEM* that utilizes thin wire as the tool electrode unlike *SEM*, where the tool is macro-sized and shaped according to the geometry to be produced in the workpiece.

In spark erosion machining (*SEM*) technology, a pulsed DC power supply is applied between tool electrode and a workpiece for spark generation. In the event of a spark discharge current flow is induced across the gap between the tool electrode and part to be machined. The energy contained in the spark discharge removes a fraction of workpiece material. A large number of these time spaced discharges between the workpiece and tool electrode causes controlled thermo-electric erosion of the workpiece material. Since, erosion is produced by electrical discharges; both tool electrode and workpiece have to be electrically conductive.

Due to the use of a thin wire as a tool, *WSEM* requires low voltage and current, high pulse frequency, longer pulse-off time and shorter pulse-on time when compared to *SEM*. Consequently, it also requires a dielectric having a low dielectric strength. De-ionized water is therefore the most commonly used dielectric due to its low viscosity and rapid cooling rate. To avoid wire breakage, there should not be any contact between the wire and workpiece during the entire process. Reduced electrode wear, lower energy consumption and independency from complicated electrode fabrication are some of the advantages of *WSEM* over *SEM* [1–4].

© Springer Science+Business Media Singapore 2017

K. Gupta and N.K. Jain, *Near-Net Shape Manufacturing of Miniature Spur Gears by Wire Spark Erosion Machining*, Materials Forming, Machining and Tribology, DOI 10.1007/978-981-10-1563-2_2

In *Micro-SEM* an electrode with micro features is used to produce its mirror image in the workpiece. This requires submicron machine movement resolution to obtain acceptable results. Similarly, in micro-WSEM a thin wire (micro size) is used to cut the workpiece that is also mounted on an accurate submicron resolution movement table. During micro-SEM the pulse generator may be controlled to produce pulses with durations ranging in length between a few nano seconds to a few micro seconds to control the extent of material volume removal [1–4].

2.1 Introduction to WSEM

The phenomenon of material erosion by electric spark was first noticed by *Joseph Priestly* in 1878, but this concept could not use for machining until 1930s. Controlled machining by electric sparks was first done by *Lazarenko* in *Russia* in 1944. In 1969, the SWISS firm '*AGIE*' produced world's first WSEM with simple features and had limitations of wire materials to be copper and brass only. Early WSEM machines were extremely slow with limited machining capabilities but as the technology matured the overall capabilities of WSEM have improved significantly to meet the requirements of various manufacturing needs. Nowadays, most of the WSEM machines are computer numerically controlled (CNC) which helps in improving the efficiency, accuracy and repeatability.

Figure 2.1 illustrates the working principle of WSEM process. A typical WSEM machine tool comprises of a main work table (called as X-Y table), an auxiliary table (U-V) and a wire drive mechanism. The workpiece is mounted and clamped on the main work table with the help of clamps and bolts. The motion along X and Y axes is controlled by means of servo-motors. A very thin wire made of brass or copper or tungsten and having diameter in the range of 0.01–0.3 mm is continuously fed from the wire feed spool and after use is collected in the waste-wire box. The wire is supported under tension, between a pair of wire guides which are disposed on both lower and upper sides of the workpiece. The wire is not reused. The dielectric is supplied through the nozzle coaxial with the wire feed system so as

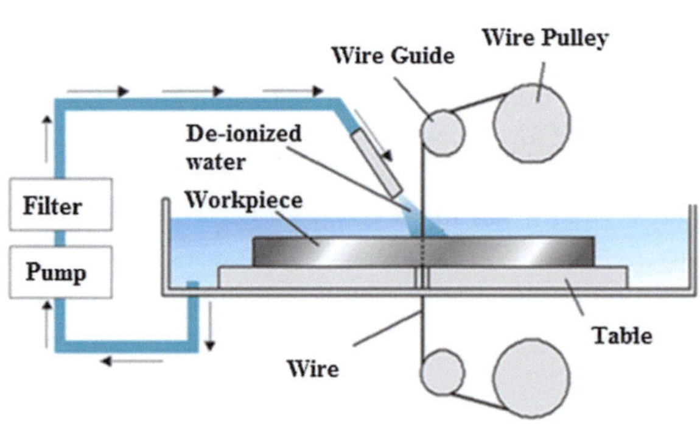

Fig. 2.1 Working principle of WSEM

to continuously flush the machining zone. A resin is used in the dielectric circulating system to maintain the conductivity of the dielectric constant. As the machining proceeds, the work table carrying the workpiece is moved along the path determined according to the geometry of the shape to be machined using the CNC program stored in the machine controller. The controller also maintains machining gap constant at the programmed value. The auxiliary U-V table is used for 3D-profiling applications.

2.2 Process Parameters of WSEM

The important parameters of WSEM are discussed below:

(a) ***Pulse-on time or Pulse duration***: The work material is eroded in WSEM during the pulse-on time when the spark occurs between wire and the workpiece. During this time, voltage is applied across the electrodes and current starts flowing. Consequently spark occurs and sustain. Longer the pulse-on time, longer the spark sustains and more is the material removal rate. However, the resulting craters are broader and deeper which result in generation of rough surface.

(b) ***Pulse-off time or Pulse interval***: It is the duration between occurrences of two consecutive sparks. The voltage is absent during this part of cycle. It is the time during which there is no power supply to the electrodes and de-ionization of dielectric takes place. Dielectric also flushes out the removed material from the machining gap during this time. Too short pulse-off time may cause wire breakage and increases the surface roughness of the machined surface due to improper removal of material from the gap, whereas too long pulse-off time increases the machining time and the forces generated by the dielectric flushing [5, 6].

Duty cycle (ratio of pulse-on time to sum of pulse-on time and pulse-off time) and *pulse frequency* (reciprocal of sum of pulse-on time and pulse-off time) are other dependent parameters describing the pulse power supply in SEM and WSEM.

(c) ***Spark Gap Voltage***: It is the reference voltage for the actual gap between the workpiece and the wire for the spark to occur between them. The machine senses the actual gap and voltage across it during the machining and tries to maintain them constant or does not allow the gap to increase to such a value that occurrence of the spark is stopped.

(d) ***Peak Current***: Peak current is the maximum value of the current passing through the electrodes for the given pulse. Increase in its value will increase the pulse discharge energy which in turn can improve the cutting rate further. But for higher value of peak current, gap conditions may become unstable.

(e) ***Wire Feed Rate***: It is the rate at which the wire is fed through the wire guides. A lower wire feed rate causes frequent wire breakage which results in poor surface finish and interruptions during the machining (i.e. less productivity) while, very high wire feed rate causes wastage of wire because it is not reused.

(f) ***Wire Tension***: It is a tensile load with which the wire is continuously fed so that it remains straight between the wire guides. Loose wire causes frequent wire breakage, dimensional and geometric inaccuracy and poor machined edge definition. Higher thickness of the workpiece requires more wire tension.

(g) ***Dielectric Pressure***: It is the pressure of the dielectric with which it flows in the machining gap for flushing the removed material. Higher dielectric pressure is required while using higher pulse power and for machining the thicker workpiece. While, low dielectric pressure is used for thin workpiece and for trim cuts.

Wire material, wire diameter, type of dielectric, its conductivity and flow rate are also important parameters which depend on the capabilities and constraints of any individual WSEM machine.

2.3 Advantages and Limitations of WSEM

Advantages

- A thin wire is used as the tool; therefore, the need of fabricating the complex tool of the complementary shape of workpiece shape is eliminated. This is the major limiting factor while using SEM. In fact, WSEM is mostly used to fabricate the complex tools for SEM.
- Any complex shape can be produced.
- No direct contact between the workpiece and the wire eliminates the mechanical stresses during machining.
- Use of significantly lower voltage and pulse-on time, as compared to SEM, also minimizes the thermal damage to the workpiece material.
- Can machine any material irrespective of its hardness, toughness or brittleness, melting points provided it is electrically conducting.
- Most of the WSEM machines are generally CNC therefore unattended machining is possible.
- Produces high quality surface finish and better dimensional accuracy.
- Generates burr-free, corrosion and wear resistant surfaces which generally do not require any post-finishing operation.

Limitations

The deeper and irregular shaped craters produced by violent spark generated at high discharge energy parameters settings and deflection of wire from its intended path known as *wire lag* are the main causes for poor surface finish, surface defects and geometrical inaccuracies in WSEMed products [6–11]. Wire lag is caused due to the impact of the mechanical forces produced by pressure of the gas bubbles, the axial forces applied to straighten the wire, the hydraulic forces induced by the dielectric flushing, the electro-static forces acting on the wire, and the electro-dynamic forces inherent to the spark generation. Nevertheless, these undesirable phenomenon up to a large extent can be controlled by optimized settings of WSEM process parameters. Other limitations of WSEM include:

- High capital cost.
- Problem of formation of recast layer particularly at higher discharge energy parameter settings.
- Not suitable for mass production.
- Not applicable to very large workpieces.

2.4 Applications of WSEM

WSEM process have extensively been using in automotive, aerospace, medical, mould, tool and die making industries. The machine's ability to operate unattended for hours or even days further increases the attractiveness of the process. Machining thick sections of material, with accuracy in dimensions and forms make this process especially valuable for the fabrication of various dies used in extrusion, stamping, powder metallurgy and injection molding. Without WSEM, the fabrication process for dies and punches requires many hours of electrodes fabrication for the conventional SEM technique, as well as many hours of manual grinding and polishing. With WSEM, the overall fabrication time is reduced greatly. Net-shape and near-net shape manufacturing of injection molding and stamping dies for gears, micro components such as micro-shafts and pipes, micro-electrodes, micro-punches, and injection nozzles are some special applications of WSEM. Difficult-to-cut materials such as superalloys, tool steels, titanium alloys, stainless steel, high speed steel (HSS), sintered carbides, metal matrix composites etc. used in automobiles, aerospace, nuclear, cutting tools, dies and moulds, and biomedical applications can be easily machined by WSEM.

2.5 Machining of Gears by Spark-Erosion Processes

Spark-erosion machining offers unique capabilities for manufacturing miniature gears. The size of the gear manufactured by *SEM* and *WSEM* depends on the size of the electrode and wire used. The path tracing capability, accuracy of the machine and appropriate process parameters affect the quality of the manufactured gear [12].

Any electrically conductive material irrespective of its hardness and melting point can be processed by *SEM/WSEM* to manufacture gears, gear cutting tools, ratchet wheels and splines. The geometric accuracy and surface finish of the gear obtained by *SEM* or *WSEM* may eliminate the need of subsequent finishing operation.

A schematic representation of gear machining by the spark erosion process is presented in Fig. 2.2. Initially a potential difference (voltage) is applied between the tool electrode/wire and the gear blank (Fig. 2.2a). The breakdown of dielectric is initiated at the closest point between the electrode and the gear blank. This increases the electric field in the gap, until it reaches the necessary value for breakdown. When the breakdown occurs the voltage falls and the current rises abruptly. The flow of current at this stage is due to the ionization of the dielectric and formation of a plasma channel between the electrode and the gear blank. The elevated current continues to further ionize the channel and a powerful magnetic field is generated (Fig. 2.2b). This magnetic field compresses the ionized channel and results in localized heating. Even with discharges of short duration, the temperature of the electrodes can rise to such an extent that the gear blank material melts locally (kinetic energy associated with the electrons are transformed into heat). The high energy density erodes a part of the material from both the electrode/wire and gear

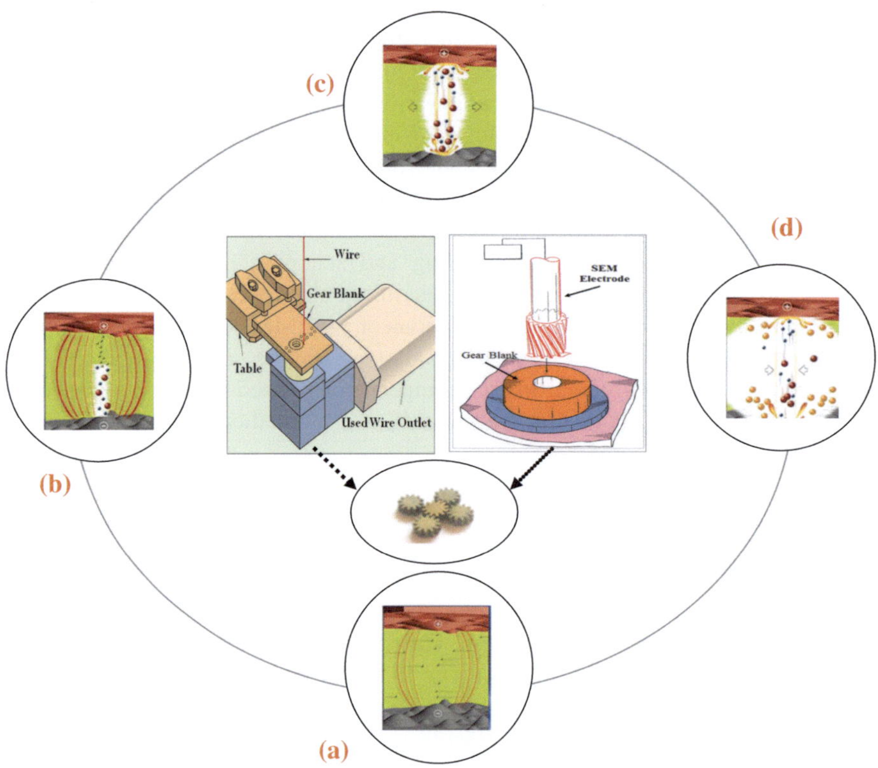

Fig. 2.2 Mechanism of material removal during machining of gears by SEM and WSEM [13]

blank by locally melting and vaporization (Fig. 2.2c). At the end of the discharge, current and voltage are shut down (Fig. 2.2d). The plasma implodes under the pressure imposed by the surrounding dielectric. Consequently, the molten metal pool is taken up into the dielectric, leaving a small crater at the gear tooth surface. This cycle is repeated until the required amount of material to be removed or the prescribed geometry is realized.

CNC programming for manufacturing of miniature gears with *SEM* is usually simply a set of commands to displace (plunge) a gear shaped electrode along the Z-axis displacement into the gear blank, whereas for WSEM, dedicated CAM software is used to define the cutting by defining gear geometry in terms of various parameters in a separate subroutine or by importing a CAD file of the appropriate geometry. Using this information, the software can generate the geometry of the gear profile to be manufactured and displays it graphically in terms of lines and arcs which represent the path of movement of the wire. The compensation for electrode size (i.e. wire diameter) and machining overcuts can also be specified. The post-processor of the software calculates all the numerical information about the movement of the wire and workpiece (gear blank) table in terms of G and M codes. The same basic procedure is followed to machine gears of any size, specification and electrically conductive material either by *SEM* and *WSEM*.

2.6 Spark Erosion Machining of Miniature Gears State-of-the-Art

The perceived improvement in geometric accuracy, quality finish and good surface integrity of products made by Spark-erosion machining (*SEM*) and its variants i.e. WSEM, micro-SEM and micro-WSEM have been the driving force for researchers and scientists to use these processes to manufacture parts for MEMS and other miniaturized-devices [13–17].

Various investigations have reported on the manufacturing of miniature gears using SEM and WSEM. These are summarized and presented in Table 2.1.

2.6.1 Spark-Erosion Machining of Gears

Meticulously done extensive literature review on miniature gear manufacturing by SEM based processes found few articles on *SEM* (or EDM) of miniature gears and all are based on *micro-SEM* of gears.

A micro-planetary gear system (0.03 mm) for a chain-type self-propelled micro-machine used in power plants was fabricated from SKS3 tool steel and WC–Ni–Cr super hard alloy by Takeuchi et al. [18] using *micro-SEM*. This system

Table 2.1 Summary of past work on manufacturing of miniature gears by spark-erosion-based processes [13]

Sr. No	Researcher	Gear type	Specification	Gear material	Process used	Findings
1.	Takeuchi et al. [18]	Micro spur gear	Module: 30 μm	SKS3 tool steel and WC–Ni–Cr cermets	Micro-SEM	Good in torque transmission performance; Dimensional variation: 0.4 %
2.	Takahata et al. [19]	Micro spur gear	Outside diameter: 200 μm; Face width: 1000 μm	WC–Co super-hard alloy	Micro-SEM	Variation in outside diameter: 4 μm
3.	Takahata and Gianchandani [20]	Micro spur gear	Outside diameter: 300 μm; Face width: 70 μm	WC–Co super-hard alloy	Micro-SEM	Fabrication time: 15 min
4.	Hori and Murata [21]	Micro spur gear	Module: 24 μm; Outside diameter: 0.280 mm; Number of teeth: 9; Face width: 0.3 mm	–	Micro-WSEM	Profile Error: 1 μm; Uniform profile with no undercutting at root area
5.	Suzumori and Hori [22]	Meso spur gear	Module: 63 μm; Outside diameter: 9 mm; Number of teeth: 95–96; Face width: 3 mm	Steel	Micro-WSEM	Satisfactory performance of the motor (equipped with these gears) under high load conditions
6.	Benavides et al. [23]	Meso ratchet wheel	Outside diameter: 6.4 mm; Face width: 0.88 mm	304 SS; austenitic stainless; beryllium copper and titanium	Micro-WSEM	Submicron level surface finish; good profile characteristics; minimum recast layer
7.	Schoth et al. [24]	Micro spur gear	Outside diameter: 1 mm, 0.5 mm;	X38CrMoVS_1 Steel SiSiC	WSEM	Good geometry and surface quality

(continued)

Table 2.1 (continued)

Sr. No	Researcher	Gear type	Specification	Gear material	Process used	Findings
			Number of teeth: 8; Face width: 6 mm, 10 mm			
8.	Di et al. [25]	Micro spur gear	Module: 100 μm; Number of teeth: 7; Face width: 1 mm	Stainless steel	Micro-WSEM	Accuracy: ±0.2 μm; Thickness of recast layer: 2 μm
9.	Ali and Mohammad [26]	Meso spur gear	Outside diameter: 3.58 mm; Number of teeth: 17; Face width: 6 mm	Copper	WSEM	Average roughness: 1 μm; Maximum roughness: 7 μm; Dimensional variation: 1–2 %
10.	Ali et al. [27]	Meso spur gear	Outside diameter: 3.58 mm; Number of teeth: 17; Face width: 6 mm	Beryllium-copper	WSEM	Average roughness: 1.8 μm; Maximum roughness: 7 μm; Dimensional accuracy: 2–3 μm
			Outside diameter: 1.2 mm; Number of teeth: 17; Face width: 6 mm		Micro-WSEM	Average roughness: 50 nm; Dimensional accuracy: 0.1–1 μm

was used as a micro-power reducer and performed satisfactorily for 5×10^6 total rotations at input torque of 10^{-7} Nm.

Takahata et al. [19] machined a WC–Co micro-gear of high aspect ratio (i.e. 5) by *micro-SEM* with only a 4 μm variation in outside diameter which could be adequately used as a micro-mechanical processing tool. Lithography, electroplating and moulding (LIGA) was used to fabricate negative-type gear electrodes of 200 μm outside diameter in nickel (See Fig. 2.3a). The discharge gap produced a reduction of 3 μm in the outside radius of the fabricated gear. This resulted in a micro-gear of WC–Co with an outside diameter of 194 μm, 1000 μm in length and with only a 4 μm variation in outside diameter across the length (Fig. 2.3b). It was then recommended to use as processing tool for micro-mechanical applications.

Takahata and Gianchandani [20] presented a new method referred to as 'batch mode micro-SEM' for precision fabrication of complex patterns of gears simultaneously. A batch of copper electrodes of 10 μm wall thickness and 300 μm height was fabricated through LIGA. These electrodes were then used as tools in

Fig. 2.3 **a** Array of negative-type nickel electrodes fabricated by LIGA process [19]. **b** Micro-gear manufactured using LIGA fabricated nickel electrode by *micro-SEM* [19]

micro-SEM for simultaneous cutting of a batch of 36 WC–Co super-hard alloy micro-gears having a 300 μm outside diameter and 70 μm thickness. The successful production of this batch of gears in 15 min was reported.

Despite the lack of available literature as regards to spark-erosion machining of miniature gears, as demonstrated above, it may be concluded that this is a practical fabrication technique for realizing high-precision mechanical systems with high robustness.

2.6.2 Wire Spark-Erosion Machining of Gears

The amount of literature available on *WSEM* (or WEDM) and *micro-WSEM* of miniature gears indicates the increased interest of researchers to explore this specific process in greater detail. Continued published work dating from the late nineties to the present amply demonstrates the capability and perceived superiority of wire spark-erosion machining for manufacturing of miniature gears. The following paragraphs briefly introduce the available literature on the manufacture of miniature gears by WSEM.

In probably the earliest published study a micro-involute spur gear of 0.28 mm outside diameter was fabricated by Hori and Murata [21] using micro wire spark-erosion machining with a tungsten wire of 25 μm diameters. The machining resulted in a burr-free uniform involute tooth profile with less than 1 μm profile error as demonstrated in the post-fabrication metrological testing and scanned electron micrograph study.

A prototype wobble motor equipped with stator and rotor having composite (involute and arc) teeth profiles for high torque and low load applications was developed by Suzumori and Hori [22]. The rotor and the stator had 95 and 96 teeth respectively. The prototype was 6 mm in pitch diameter, 9 mm in outside diameter, and 3 mm thick. It was fabricated from steel by *micro-WSEM*. The motor maintained almost constant speed and full wobble motion until just stalling during performance testing, unlike a motor equipped with involute tooth profile rotor-stator.

Sandia National Laboratories in the USA has been actively involved in design, development, and fabrication and testing of micro electro mechanical systems (MEMS) for many years resulting in several published and unpublished studies on spark-erosion machining of miniature gears for MEMS and other miniaturized devices. Benavides et al. [23] from the Manufacturing Science and Technology Centre at *Sandia lab* employed *micro-WSEM* to fabricate a meso-sized ratchet wheel of different materials (i.e. 304L stainless steel, nitronic 60, austenitic stainless, beryllium copper, and titanium). A submicron level surface finish, burr-less edges and profiles, minimum recast layer and consistent micro-geometry were achieved. A total of seven test parts were fabricated for their work: two each of 304L SS, nitronic-600 annealed stainless steel and titanium alloy and one of Beryllium Copper. The machining duration on the *micro-WSEM* machine was

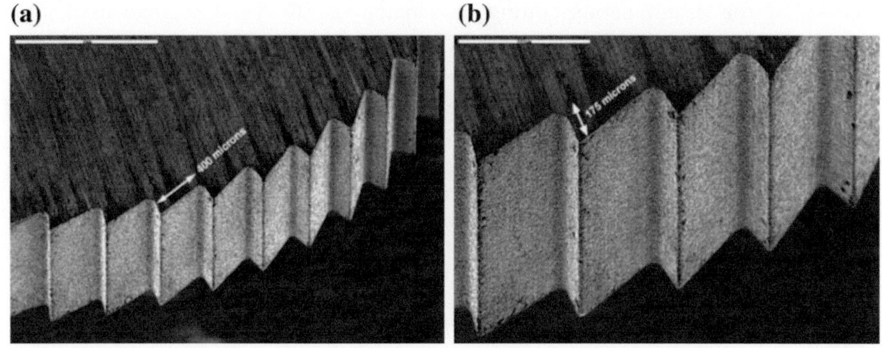

Fig. 2.4 Scanned electron micrograph of micro-WSEMed ratchet teeth in Nitronic 60 stainless steel [23]. **a** Normal view. **b** Magnified view

approximately two hours per part. The *Micro-WSEM* machine was equipped with tungsten wire of 30 μm diameter and de-ionized water as dielectric. Figure 2.4a, b depicts the scanned electron microscopic images of the ratchet teeth cut into a nitronic 60 ratchet wheel test part.

All materials displayed satisfactory edge definition and very thin recast layers (See Fig. 2.5a–d). However, Fig. 2.5c indicates that the titanium alloy ratchet wheel's features are smoothest and have minimum recast layer. Metrology investigations for the fabricated ratchet wheels (details are given in Table 2.2) indicate that the ratchet wheels of 304L stainless steel have the best profile tolerance

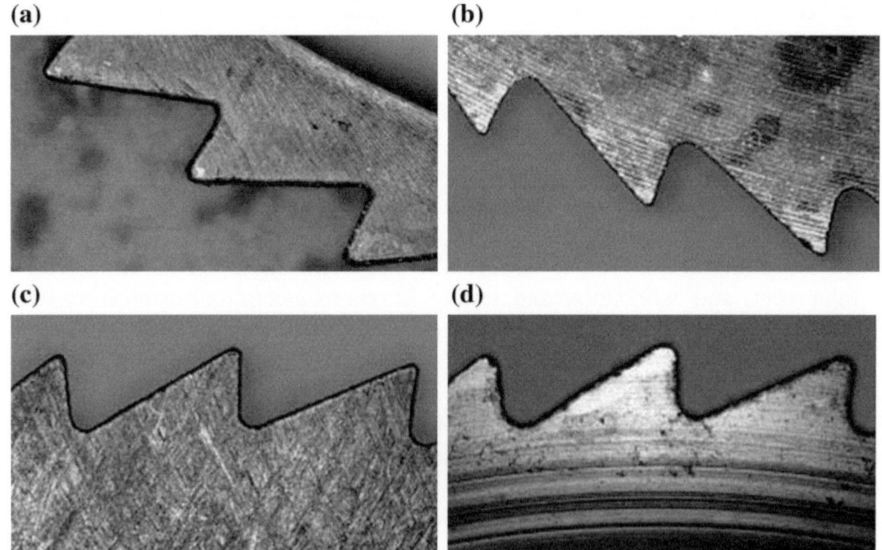

Fig. 2.5 Profile micrographs of teeth of micro-WSEMed ratchet wheels [23]. **a** 304L SS. **b** Nitronic 60 SS. **c** Titanium alloy. **d** Beryllium copper

Table 2.2 Results of a metrology investigation of fabricated ratchet wheels [23]

Material		Profile tolerance (in microns)
304L SS	Part no. 1	1.5
	Part no. 2	1.5
Nitronic 60 SS	Part no. 1	4.1
	Part no. 2	4.1
Titanium alloy	Part no. 1	2.3
	Part no. 2	2.3
Beryllium copper	Part no. 1	3.8

followed closely by the titanium, thus making them the most favorable materials for fabrication of precision meso-scale parts by *micro-WSEM*.

Schoth et al. [24] demonstrated the capability of *Micro-WSEM* fabrication of high aspect ratio 3D microstructures in selected ceramics and metals. A 30 μm tungsten wire was used to fabricate a gear wheel of X38CrMoVS_1 steel with a 1 mm outside diameter, 6 mm thickness and 8 teeth (Fig. 2.6a). The same wire was used to fabricate a ceramic (SiSiC) gear wheel (Fig. 2.6b) of 1 mm outside

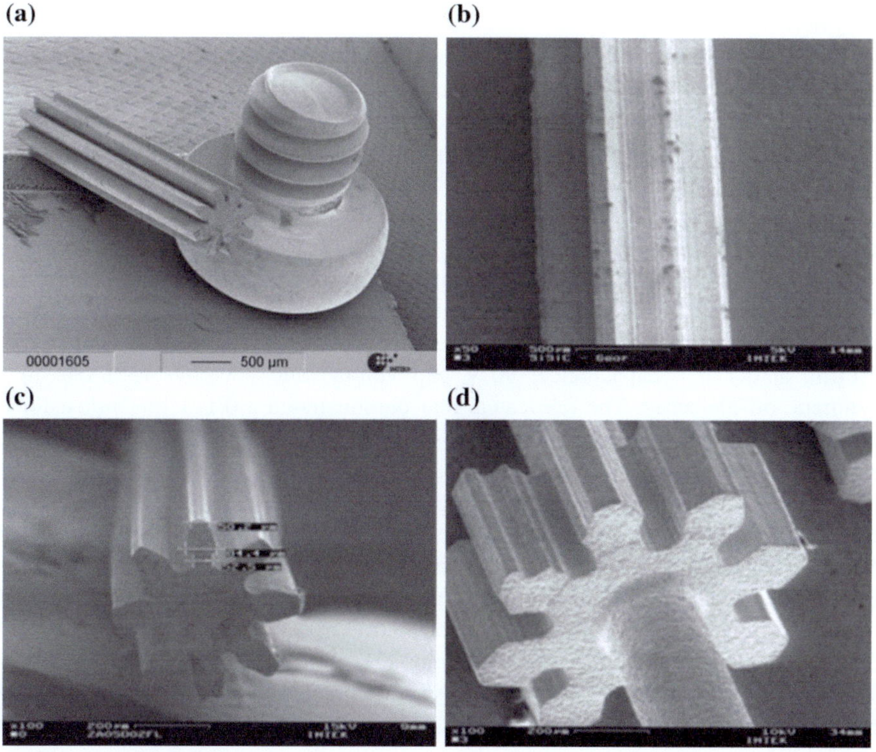

Fig. 2.6 Micrographs of teeth of micro-WSEMed gear wheels [24]

diameter, 10 mm thickness and 8 teeth. Another smaller gear with a 0.5 mm outer diameter, 6 mm height and 8 teeth was machined in X38CrMoVS_1 steel by utilizing a 20 μm tungsten wire (Fig. 2.6c). A gear wheel with integrated shaft, for ease of assembly, was also fabricated by *micro-WSEM* (Fig. 2.6d). They concluded that the electrode diameter was the most significant parameter that affected accuracy.

Di et al. [25] manufactured stainless steel micro-internal gears of 100 μm module with 7 teeth and 1 mm thickness by *micro-WSEM* using a tungsten wire of 30 μm diameter and at a speed of 20 mm/min. The best fabricated gear achieved ±0.2 μm accuracy, 0.1 μm surface roughness and 2 μm thick recast layer. Micro-forming dies of module 100 μm and 3.5 mm thickness have also been fabricated and have consequently been used successfully to form micro-gears of aluminium alloys.

In another study, Ali and Mohammed [26] machined external meso-spur gears with 3.58 mm outside diameter and 17 teeth from a 6 mm thick copper blank with *WSEM* utilizing brass wire of 100 μm diameter. They obtained a 1.4 μm average roughness, 7 μm maximum roughness and only 1–2 % average dimensional variation. Subsequent microstructural investigation and analysis showed the existence of shallow craters and other irregularities on the gear teeth surfaces. Consequently, lower discharge energy parameter settings were recommended for better surface integrity.

Thereafter, Ali et al. [27] compared two variants of SEM namely conventional *WSEM* and *micro-WSEM* for manufacturing of micro-spur gears. In the conventional *WSEM* process, a best set of parameters was selected during meso-fabrication of rectangular plate of Be–Cu alloy. Using this parameter settings (1 A current, 5 V gap voltage, 6 μs pulse-on time, and 5 μs pulse-off time) a meso-gear (17 teeth, 3.5 mm outside diameter and 6 mm thickness) in beryllium-copper alloy was fabricated utilizing brass wire of 100 μm diameter with de-ionized water as dielectric. This gear had 2–3 μm dimensional accuracy, 1.8 μm average roughness and 7 μm maximum roughness. Subsequently a meso-gear (17 teeth and 1.2 mm outside diameter) was machined by *micro-WSEM* using a wire with diameter of 70 μm, at 0.1 n F capacitance; 90 V gap voltage and 3.8 μm/s feed rate in a synthetic oil dielectric. The fabricated gear demonstrated a 0.1–1 μm dimensional accuracy and 50 nm average roughness. Figures 2.7 and 2.8 depict various micrographic views of both the meso-gears.

A micrograph surface study of both gears indicated a crack-free surface structure. A comparison of the dimensional variation and surface roughness revealed *Micro-WSEM* as the superior technique albeit with the disadvantage of being slower than conventional *WSEM*. During experimentation they also observed the increased material removal rate and subsequent deterioration in surface quality with increased discharge current, gap voltage (in conventional WSEM) and capacitance (in *micro-WSEM*). It was therefore unsurprisingly concluded that low discharge energy parameters have to be used in *WSEM* for superior geometrical features and surface finish [27].

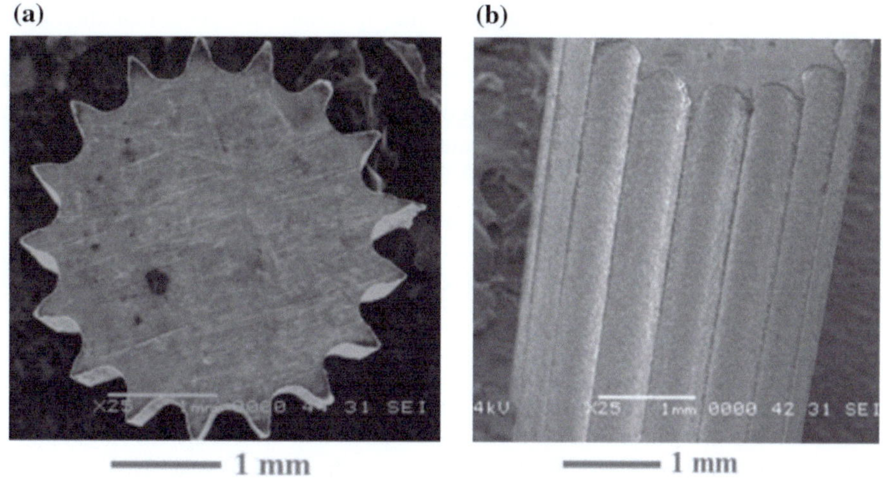

Fig. 2.7 Meso-spur gear machined by conventional *WSEM* [27]. **a** Top view. **b** Isometric view

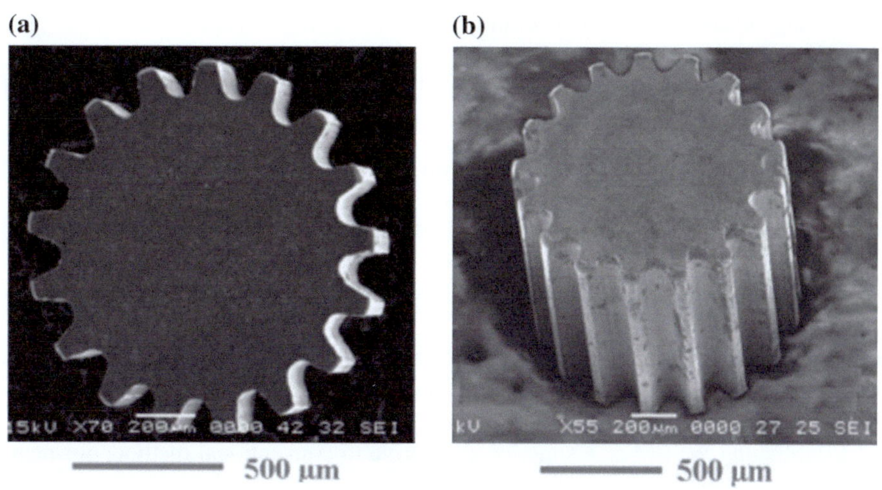

Fig. 2.8 Meso-spur gear machined by *micro-WSEM* [27]. **a** Top view. **b** Isometric view

Critical review based on spark-erosion machining of gears has shown that there is hardly any literature available on machining of macro-gears by spark erosion based machining processes except for the attempt made by Talon et al. [28], in which a macro size spur gear (outside diameter 28 mm; module 2 mm; 12 tooth; 5 mm face width) of important aerospace titanium alloy i.e. Ti–6Al–4V was precisely fabricated in 50 min using WSEM. Post fabrication metrological inspection of this gear by gear tooth Vernier calliper and coordinate measuring machine

(CMM) concluded that a lower dimensional variation and good manufacturing quality (i.e. ISO-7) were obtained.

It became abundantly clear that in most cases only the capability of *SEM/WSEM* to fabricate the miniature gears was of concern and the parameters on quality such as micro-geometry, surface finish and integrity; and productivity were addressed in passing only.

2.7 Conclusions and Scope of Spark Erosion Machining of Miniature Gears

The review of past work as presented in the forgoing sections on *SEM* and *WSEM* of miniature gears indicates a clear lack of detailed and systematic investigation of how the different process parameters affects the quality aspects (especially micro-geometry) of miniature gears. Some major conclusions in terms of research gaps include lack of work on WSEM of meso-sized gears; lack of investigation on the effect of spark erosion process on micro-geometry of gears; and lack of focus on optimizing spark erosion machining process for high quality miniature gear manufacturing.

Based on the limitations of past work, the following points define the scope of research work discussed in the subsequent chapters of this book:

- Exploring the capability of WSEM for manufacturing high quality miniature gears.
- Investigating and analysing the effects of WSEM parameters on micro-geometry, surface roughness and surface integrity of miniature gears.
- Improving the productivity of WSEM while manufacturing high quality miniature gears.
- Modelling and optimization of WSEM process for high quality and productivity.
- Comparative study of gear hobbing and WSEM to manufacture high quality miniature gears.
- Establishing WSEM as a superior alternative to conventional methods for near-net shape manufacturing of miniature gears.

References

1. Gupta K, Jain NK, Laubscher RF (2016) Chapter-1: Overview of hybrid machining processes, in Hybrid Machining Processes. Springer International Publishing Co. Ltd AG, Cham
2. McGeough JA (1988) Advanced methods of machining. Chapman and Hall Ltd, London
3. Jain VK (2002) Advanced machining processes. Allied Publishers, New Delhi
4. Ho KH, Newman ST, Rahimifard S, Allen RD (2004) Start of the art in wire electrical discharge machining. Int J Mach Tool Manuf 44(12–13):1247–1259

5. Rajurkar KP, Wang WM, Lindsay RP (1991) On-line monitor and control for wire breakage in WSEM. Ann CIRP 40(1):219–222
6. Okada A, Konishi T, Okamoto Y, Kurihara H (2015) Wire breakage and deflection caused by nozzle jet flushing in wire EDM. Ann CIRP 64:233–236
7. Arunachalam C, Aulia M, Bozkurt B, Eubank PT (2001) Wire vibration, bowing, and breakage in wire electrical discharge machining. J Appl Phys 89(8):4255–4262
8. Liao YS, Huang JT, Chen YH (2004) A study to achieve fine surface finish in Wire-SEM. J Mater Process Technol 149:165–171
9. Mingqi L, Minghui L, Guangyao X (2005) Study on the variations of form and position of the wire electrode in WEDM-HS. Int J Adv Manuf Technol 25:929–934
10. Puri AB, Bhattacharyya B (2003) An analysis and optimization of the geometrical inaccuracy due to wire lag phenomenon in WEDM. Int J Mach Tool Manuf 43:151–159
11. Sarkar S, Sekh M, Bhattacharyya B (2011) A novel method of determination of wire lag for enhanced profile accuracy in WEDM. Precis Eng 35:339–347
12. Gupta K, Jain NK (2014) On surface integrity of miniature spur gears manufactured by wire electrical discharge machining. Int J Adv Manuf Technol 72(9–12):1735–1745
13. Gupta K, Jain NK, Laubscher RF (2015) Spark-erosion machining of miniature gears: a critical review. Int J Adv Manuf Technol 80(9–12):1863–1877
14. Bhuiyan M, Shihab B (2014) Development of copper based miniature electrostatic actuator using WEDM with low actuation voltage. Microsyst Technol. doi:10.1007/s00542-014-2360-z
15. Yan J, Uchida K, Yoshihara N, Kuriyagawa T (2009) Fabrication of micro end mills by wire-EDM and some micro cutting tests. J Micromech Microeng 19(025004):9
16. Dokania AK, Pelle M, Kruit P (2008) Fabrication of miniaturized schottky emitter by wire electrical discharge machining (WEDM). Microelect Eng 85:1031–1034
17. Uhlmann E, Piltz S, Doll U (2005) Machining of micro/miniature dies and moulds by electrical discharge machining-recent development. J Mater Process Technol 167:488–493
18. Takeuchi H, Nakamura K, Shimizu N, Shibaike N (2000) Optimization of mechanical interface for a practical micro-reducer. In: Proceedings of 13th IEEE international conference on micro electro mechanical systems. Miyazaki, Japan
19. Takhata K, Shibaike N, Guckel H (2000) High-aspect-ratio WC-Co microstructure produced by the combination of LIGA and micro-EDM. Microsyst Technol 6:175–178
20. Takhata K, Gianchandani YB (2002) Batch mode micro-electro discharge machining. J Micro Mech Syst 11(2):102–110
21. Hori K, Murata Y (1994) Wire electrical discharge machining of micro-involute gears. Transact Japan Soc Mech Eng Ser C 60(579):3957–3962
22. Suzumori K, Hori K (1997) Micro electrostatic wobble motor with toothed electrodes. In: Proceedings of 10th international workshop on micro electro mechanical systems. IEEE, Nagoya, Japan, pp 227–232
23. Benavides GL et al (2002) High aspect ratio meso-scale parts enabled by wire micro-EDM. Microsyst Technol 8:395–401
24. Schoth A, Forster R, Menz W (2005) Micro wire EDM for high aspect ratio 3D microstructuring of ceramics and metals. Microsyst Technol 11:250–253
25. Di S, Haung R, Chi G (2006) Study on micro-machining by micro-WSEM. In: Proceedings of the 1st international IEEE conference on nano/micro engineered and molecular systems. Zhuhai, China
26. Ali MY, Mohammad AS (2008) Experimental study of conventional WEDM for micro-fabrication. Mater Manuf Process 23(7):641–645
27. Ali MY et al (2010) Comparative study of conventional and micro-WEDM based on machining of meso/micro sized spur gear. Int J Precis Eng Manuf 11(5):779–784
28. Talon JLH et al (2010) Manufacturing of a spur tooth gear in Ti-6Al-4V by electrical discharge. Comp Aid Design 42:221–230

Chapter 3
Planning, Design and Details of Experimental Investigation

3.1 Introduction to Design of Experiments

Experimentation is required to investigate the changes that may be observed in the responses due to the changes made in the input parameters of a process or system, so as to improve its performance by identifying and studying the reasons for it. If the effect of a parameter depends on the level of other parameter then two parameters are said to have interaction. Principles of design of experiments (DOE) help in selecting optimal product or process parameters. The advantages of using DOE approach are as follows [1]:

- It reduces the cost and time of experimentation by minimizing the number of experiments required for gathering the same amount of the information as compared to conducting the experiments randomly.
- It helps in identifying the optimum values of process parameters.
- It helps in modeling the performance measures of the process in terms of the process parameters and may be their interactions.

Any DOE approach should incorporate following features of replication, randomization, and blocking [2]:

- **Replication**: It involves repetition of the experiment under similar conditions to obtain an estimate of the experimental error (i.e. variation in the responses) under identically controlled conditions. The experimental error forms a basis for determining whether differences in the statistics found from the observations are significant. It helps to determine the significant input parameters and interactions influencing the responses; and increases the precision by reducing the standard deviation of the mean in case the sample mean is used to estimate the effect of a factor.

© Springer Science+Business Media Singapore 2017
K. Gupta and N.K. Jain, *Near-Net Shape Manufacturing of Miniature Spur Gears by Wire Spark Erosion Machining*, Materials Forming, Machining and Tribology, DOI 10.1007/978-981-10-1563-2_3

- **Randomization**: It assigns the input parameters to the experimental units in such a way that every unit has an equal chance of being assigned to any parameter in order to eliminate bias and to ensure that no particular parameter is favored. Randomization averages out the effect of uncontrolled factors or noise.
- **Blocking**: In blocking the similar experimental units are arranged into groups, or blocks. Variability of the response function within a block can then be attributed to the differences in the input parameters because the impact of the other extraneous variable has been minimized. Input parameters are assigned at random to the units within each block, which results in reduction of variability between block and increase in the precision of the experiment.

All the DOE approaches can be classified into two categories namely (i) *Classical statistical DOE*, and (ii) *robust DOE*. Both classical statistical and robust DOE approaches use the basic principles of planning the experiments and data analysis developed by R.A. Fisher in the 1920s. Classical statistical DOE approaches include completely randomized design (CRD), randomized block design (RBD), Latin square design (LSD), Graeco-Latin square design (GLSD), response surface methodology (RSM) which are primarily used to derive a mathematical equation relating the mean response to the levels of the input parameters assuming that the variance of the response remains constant for all the levels of the input parameters. While, the robust DOE approach (*Taguchi* method) focuses on determining the effects of the input parameters on the process performance. It capitalizes on the change in the variance and tries to minimize the variance by changing the levels of the input parameters. In the robust DOE, the accurate modeling of the mean response is not as important as finding the input parameter levels that optimizes the process performance [3, 4]. All the DOE approaches can also be classified as full factorial or fractional factorial approach. The *full factorial* DOE approach uses all the possible combinations of the input process parameters for the experimentation which ensure that optimum response lies within the domain of experimental results but, this increases the total number of experiments (i.e. *total number of the experiments required* $= \prod_{i=1}^{N} n_i$; *where* n_i *is the number of levels for* ith *parameters*) significantly with increase either in the input process parameters and/or their levels consequently increasing the cost and time required for the experimentation. While, the *fractional factorial* DOE approach scientifically reduces the total number of experiments required without compromising the information content of the experimental results and at same time it ensures that the optimum solution lies within the domain of results of the planned experiments.

One important fractional factorial DOE approach is *Taguchi* technique which is based on the concept of robustness. It uses orthogonal arrays for planning the experiments to study several decision variables simultaneously. But, *Taguchi* method possesses several limitations as pointed out by several critics [2, 4], some of them are:

1. Taguchi method attempts to identify the factors that most influence variability of the performance measure, it does not focus on the reason why this happens. It tries to minimize the variability by maximizing the signal-to-noise (S/N) ratio. Taguchi has proposed more than 60 signal-to-noise ratios for particular applications. But, maximizing S/N ratio does not guarantee reduction in the variability. In fact in certain situations, reduction of variability could overshadow the closeness to the target value of the responses.
2. Taguchi's DOE while using S/N ratio as measure of the performance implicitly assumes that the standard deviation of the response is linearly proportional to its mean. Maximization of the S/N ratio occurs over the levels of the parameters considered and not over the entire feasible region.
3. S/N ratio used in the Taguchi method do reveal the location effects due to the mean but, it may not be effective in identifying the dispersion effects because location and dispersion effects are confounded in the S/N ratio.
4. Taguchi method prefers using three or more levels of the parameters to investigate non-linear effects (i.e. curvature) rather than investigating the potential interactions. There are three types of interactions possible namely (i) among the design parameters; (ii) among the design parameters and noise parameters; and (iii) among the noise parameters. Taguchi method though recognizes the interactions between the design parameters but downplays their importance relative to the main effects.
5. Taguchi's procedure for computing marginal averages may not in general identify the optimum combinations of the parameters because it ignores the interaction effects between the parameters. Use of marginal averages is analogues to the one-factor at a time approach.
6. Taguchi method focuses on whether the variability is most influenced by the main effects and their interactions instead of identifying the components cause the variability to happen and the nature in which they contribute to the variability as attempts by the classical fractional factorial DOE approach such as RSM.
7. The linear graphs used in the Taguchi Method are heuristically developed and the main effects are confounded with two-way interactions.

3.1.1 Response Surface Methodology (RSM)

The main experiments of the present work are designed accordingly to the *Box-Behnken* Design (BBD) of RSM. G.E.P. Box and K.B. Wilson developed RSM in 1951. RSM is a collection of mathematical and statistical technique that is useful for modeling and analysis of problems in which a response of interest is influenced by several variables and objective is to optimize this response [4]. It attempts to determine the shape of the response function and its sensitivity to the parameter design factors. If a response '*Y*' is a function of variable parameters X_1, X_2, ..., *Xn* as represented by Eq. 3.1; which is to be maximized

$$Y = f(X_1, X_2, \ldots, X_n) + \epsilon \tag{3.1}$$

Here, ε represents the error observed in the values of the response. If expected response is denoted by *E*(*Y*) as defined below:

$$E(Y) = f(X_1, X_2, \ldots, X_3) = \eta \tag{3.2}$$

Then the surface represented by $\eta = f(X_1, X_2, X_3)$ is called a response surface.

The RSM methods facilitate the improvements in product and process designs by understanding the nature of the response function and thereby the process. Improved product and process designs are thus facilitated. In most of the RSM problems, the form of the relationship between the response and the independent variables is unknown. RSM is a sequential procedure. The *first step* is to find a suitable approximation for the true functional relationship between response and the independent variables. Usually, a low-order polynomial in some region of the independent variables is used. If the response can be well modeled by a linear function of the independent variables, then the approximating function is the *first-order model*. If the response has non-linear dependence on the independent variables i.e. curvature is present in the response surface, then a polynomial of higher degree (generally the *second-order model*) should be used. The *second step* is to estimate the constants used in the approximating polynomials using the *least square* method of curve fitting. The *third step* is to do response surface analysis using the fitted surface. If the fitted surface is an adequate approximation of the actual response function then analysis of the fitted surface will be approximately equivalent to analysis of the actual system. The eventual objective of RSM is to determine the optimum values of the variable parameters to optimize the response(s) or to identify the region of the variable parameters in which the response(s) is/are optimized.

A good response surface design should have following desirable properties:

- Reasonable distribution of data points throughout region of interest.
- Requires minimum number of experimental runs.
- Should not require too many levels of the independent variables.
- Allows experiments to be performed in blocks.
- Model adequacy including lack of fit.
- Precise estimates of the parameters used in the models.
- Simplicity in calculation of the model parameters.
- Reasonable robustness against the outliners or the missing values.
- Sequential construction of higher order designs from simpler designs.
- Good profile of the prediction variance throughout the experimental region.
- Good graphical analysis through simple data patterns.

- Good behavior when errors in settings of input variables occur.
- Internal estimate of error.
- Constant variance check.

Mainly two DOE approaches exist under RSM: Central Composite Design (CCD) and Box-Behnken Design (BBD). CCD is a two level full factorial (2^k) or fractional factorial (2^{k-1}, $k = no.\ of\ factors$) design augmented by a number of centre and axial points [4]. It is generally used for the factors having five levels. It has three groups of data points: (a) factorial data points; (b) axial data points (sometimes called "star" points); and (c) centre points (Fig. 3.1). The factorial part of the design consists of all possible combinations of the extreme levels of the factors. The star points have all of the factors set to 0, the midpoint, except for one factor, which has the value of $\pm\alpha$, which is calculated in each design for both rotatability and orthogonality of blocks. Rotatability provides constant variance of the estimated response corresponding to all new observation points that are at the same distance from the centre point of the design. The experimenter can choose between these values or enter a different one. Another position for the star points is at the face of the cube portion on the design i.e. $\alpha = 1$. This design is commonly referred to as a *face-centered* central composite design (FC-CCD). It requires only three levels for each factor. Centre points, as implied by the name, are points with all levels set to coded level 0 i.e. the midpoint for each factor range. These points are usually repeated 4–6 times to get a good estimate of experimental error (pure error).

Box-Behnken Design (BBD) was devised by G.E.P. Box and D. Behnken in 1960. Box and Behnken have proposed some three level designs for fitting the response surfaces. These designs are formed by combining 2^k factorials with incomplete block designs. The advantages of BBD include the fact that it is

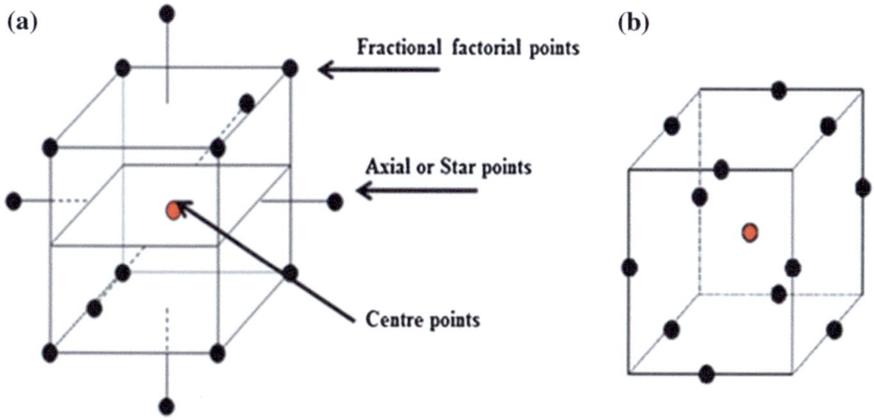

Fig. 3.1 Layout of **a** CCD; and **b** BBD for three factors

spherical design and requires factors to be run at only three levels. The designs are also rotatable or nearly rotatable. Box-Behnken approach generally used to perform non-sequential experiments i.e. experiments are performed only once. This approach requires fewer data points, and which are less expensive to run as compared to CCD with same number of the factors. The experimental combinations are at the midpoints of edges of the space defined by the ranges of the factors and at the center (Fig. 3.1). This approach does not contain any point at the vertices of region defined by the factors which could be advantageous when the points on the corners of the cube represent factor level combinations (high levels) that are prohibitively expensive or impossible to test because of physical process constraints. Figure 3.2 depicts the experimental design based on BBD approach for three factors each having three levels. It is not possible to present the geometry of the experimental design in cubic form on paper based on BBD for four factors each at three levels. Table 3.1 presents the number of the experimental runs for different number of factors in the BBD approach.

Some of BBD designs also provide *orthogonal blocking*. Thus, if there is a need for separate runs into blocks for the BBD, then designs are available that allow blocks to be used in such a way that the estimation of the regression parameters for the factor effects are not affected by the blocks. In other words, in these designs the block effects are orthogonal to the other factor effects. Table 3.2 presents the comparison of the BBD and FC-CCD approaches of RSM with the full factorial design (FFD) for the three-level factors.

The following sections provide the details of the DOE used during each stage of the present experimental work.

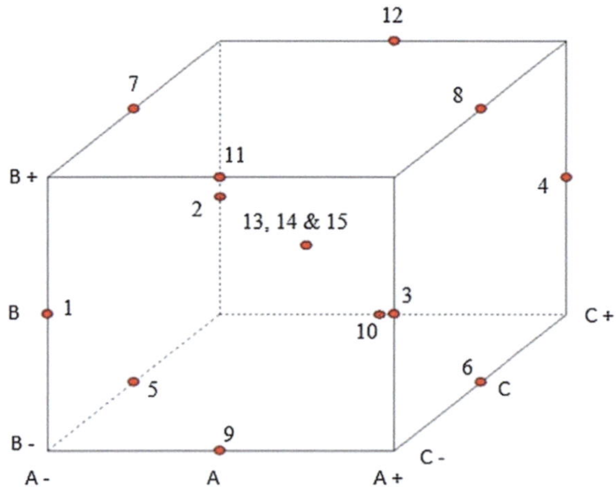

Fig. 3.2 Geometric representation of BBD for three factors having three-levels

Table 3.1 Various Box-Behnken designs for different number of factors

No. of factors	No. of factors varied in each block	No. of blocks	No. of runs with one centre point	No. of runs with extra centre points	No. of coefficients in quadratic model
3	2	3	13	15, 17	10
4	2	6	25	27, 29	15
5	2	10	41	46	21
6	3	6	49	54	28
7	3	7	57	62	36
8	4	14	113	120	45
9	3	15	121	130	55
10	4	10	161	170	66
11	5	11	177	188	78
12	4	12	193	204	91
16	4	24	385	396	153

Table 3.2 Comparison of full factorial design (FFD) with FC-CCD and BBD approaches of RSM for the three-level factors

No. of factors	Number of experimental runs required by		
	Full Factorial Design (FFD)	FC-CCD	BBD
2	9	13	NA
3	27	20	17
4	81	30	29
5	243	50	46
6	729	86	54
7	2187	148	62

3.2 Experimental Plan for Wire Spark Erosion Machining of Miniature Gears

The experimental investigations in the present work are planned in four different stages namely: (i) trial experiments, (ii) pilot experiments, (iii) main experiments, and (iv) confirmation experiments.

Table 3.3 presents the summary of objectives, details of fixed and variable WSEM parameters, measures of process performance or responses, and methodology of DOE for the pilot, main and confirmation experiments. This table also contains the specifications of the miniature gear selected to fabricate by WSEM. The DOE approach for the each experimental stage was decided keeping in view its objectives, number of the variable parameters involved and their levels, and the constraints on the experimental resources.

Table 3.3 Summary of experimental details for different stages

Experimentation stage	Objectives	Input parameters with levels	Responses	DOE approach (no. of experiments)
Pilot	• To analyze the behavior of micro-geometry parameters with WEDM parameters • To bracket the range of WEDM parameters for further investigations • To analyze the surface integrity of miniature gears	1. Voltage (V): 5–10–15–20–25	Total profile error (μm)	One factor at-a-time 23 Experiments
		2. Pulse-on time (μs): 0.6–0.8–1–1.2–1.4	Accumulated pitch error (μm)	
		3. Pulse-off time (μs): 90–130–170–210–250	Average roughness (μm)	
		4. Wire feed rate (m/min): 3–6–9–12–15	Maximum roughness (μm)	
		5. Cutting speed (%): 50–75–100	Microstructure	
			Micro-hardness	
Main	• To analyze the effect of WEDM parameters and interactions between them on the responses • To further facilitate the optimization	1. Voltage (V): 5–10–15	Total profile error (μm)	Box-Behnken 29 Experiments
		2. Pulse-on time (μs): 0.6–0.8–1	Accumulated pitch error (μm)	
		3. Pulse-off time (μs): 90–130–170	Avgerage roughness (μm)	
		4. Wire feed rate (m/min): 9–12–15	Maximum roughness (μm)	
			Gear cutting rate (mm^3/min)	

(continued)

Table 3.3 (continued)

Experimentation stage	Objectives	Input parameters with levels	Responses	DOE approach (no. of experiments)
Confirmation	• To validate the optimum results predicted by desirability analysis	Optimized WEDM parameters	Total profile error (μm)	9 Experiments
			Accumulated pitch error (μm)	
			Average roughness (μm)	
			Maximum roughness (μm)	
			Gear cutting rate (mm^3/min)	

Fixed parameters Wire material: brass; Wire diameter: 0.25 mm; Wire tension: 1200 grams, Dielectric: de-ionized water; Dielectric conductivity: 20 μS/cm; Dielectric pressure: 7 kg/cm^2

Miniature gear specifications Material: ASTM 858 brass; Profile: involute; Type: external spur gear; Pressure angle: 20°; Module: 0.7 mm; Outside diameter: 9.8 mm; Number of teeth: 12; Face width: 5 mm

3.2.1 Trial Experiments

The WSEM machine tool used in the present work had wide range of process parameters to choose from. Since, no reference about past work was available on manufacturing of high quality meso-spur gears made of brass; therefore, bracketing the feasible working range of WSEM parameters which are suitable for manufacturing the meso-gears from *brass* was required for any further experimental investigations. Gear material used in the present work is 'ASTM 858' brass (refer Appendix A for composition evaluated by optical emission spectrometry).

Consequently, total forty trial experiments were conducted by varying the WSEM parameters in their entire range available on the machine tool i.e. 1–100 V for the voltage 'V', 0.1–1.65 μs for the pulse-on time 'T_{on}', 10–260 μs for the pulse-off time 'T_{off}', and 1–15 m/min for the wire feed rate. In these experiments, simple rectangular shaped strips were cut from a rectangular brass plate of 5 mm thickness. Wire breakage frequency, roughness of the machine surface and cutting speed were investigated as the responses.

3.2.2 Pilot Experiments

Total twenty three pilot experiments were designed and conducted using one factor at-a-time approach (OFAT). The experiments designed using OFAT are regarded as easier to implement, more easily understood, economical and are better than trial and error approach. In spite of the certain limitations of this approach such as not providing efficient estimates of the effects and adequate information about interactions, this approach was used for the pilot experiments because aim of the pilot experiments was only to further narrow down the ranges of the WSEM parameters while, the detailed study of the effects of parameters and their interactions was taken care of in the main experiments. The pilot experiments were conducted by varying the four WSEM parameters (V, T_{on}, T_{off} and W) at five levels each and at three levels for the cutting speed (F) as mentioned in Table 3.3 considering responses as the micro-geometry parameters of miniature gears (i.e. total profile error 'F_a' and accumulated pitch error 'F_p'), surface roughness (i.e. average surface roughness 'R_a' and maximum surface roughness 'R_{max}') and WSEM productivity (i.e. gear cutting rate). The levels of the parameters were chosen on the basis of trial experiments. Five out of twenty three experiments had same set of parameters; therefore only nineteen experiments were conducted in order to get clear idea to bracket the parameter ranges.

3.2.3 Main Experiments

Various fractional factorial DOE approaches were compared to facilitate the selection for the main experiments. On the basis of this comparison, BBD approach of RSM was found to be the most suitable for the main experiments because of having certain advantages such as (i) BBD requires fewer design points which are less expensive to run; (ii) It does not retain any parametric combination having extreme values of the parameters; (iii) It provides strong coefficient estimates near the centre of the space bound by the parameter ranges where the presumed optimum lies; and (iv) it is rotatable or nearly rotatable [2, 4].

Twenty nine experiments were conducted based on *BBD* approach to investigate the effects of four WSEM parameters (V, T_{on}, T_{off} and W) on the F_a, F_p, R_a, R_{max} and *GCR*. This contained 24 side points and 5 replications of the center points. Each experiment (out of 29) was replicated twice, therefore total 58 miniature gears were manufactured in the main experiments. During main experiments WSEM parameters (V, T_{on}, T_{off} and W) were varied at three levels each but, cutting speed was fixed at its highest value (Table 3.3). Main experiments were aimed to: (i) find the significant WSEM parameters and their interactions affecting the responses; (ii) investigate their effects on the responses; and (iii) optimize WSEM parameters for the optimum quality of the miniature gears (i.e. minimum micro-geometry errors and surface roughness) with enhanced productivity of WSEM. Table 3.4 presents the 29 experimental combinations designed by BBD approach for the main experiments.

3.2.4 Confirmation Experiments

Nine confirmation experiments were performed to validate the optimum results given by (i) the regression models based on the results of the main experiments and integrated with the desirability functions; and (ii) the results of multi-objective optimization using back propagation neural network (BPNN) integrated genetic algorithms (GA). The nine confirmation experiments included 5 experiments for single-objective optimization of each response individually (one each for F_a; F_p; R_a; R_{max} and *GCR*); one experiment is for optimization of F_a and F_p simultaneously, and one for R_a and R_{max} simultaneously; and two experiments for multi-objective optimization of all the responses simultaneously (one for optimization by RSM-based desirability analysis and one for the optimization by the BPNN-GA approach).

Table 3.4 Experimental combinations of WSEM parameters with coded and actual values of the parameter levels designed by BBD for main experiments

Expt. No.	Variable input parameters (coded and actual values)							
	V (Volts)		T_{on} (μs)		T_{off} (μs)		W (m/min)	
1	(1)	15	(0)	0.8	(0)	130	(1)	15
2	(0)	10	(0)	0.8	(−1)	90	(−1)	9
3	(−1)	5	(1)	1.0	(0)	130	(0)	12
4	(0)	10	(0)	0.8	(0)	130	(0)	12
5	(1)	15	(−1)	0.6	(0)	130	(0)	12
6	(−1)	5	(0)	0.8	(0)	130	(−1)	9
7	(−1)	5	(0)	0.8	(1)	170	(0)	12
8	(0)	10	(1)	1.0	(0)	130	(−1)	9
9	(1)	15	(0)	0.8	(0)	130	(−1)	9
10	(0)	10	(0)	0.8	(1)	170	(1)	15
11	(1)	15	(0)	0.8	(1)	170	(0)	12
12	(1)	15	(1)	1.0	(0)	130	(0)	12
13	(0)	10	(0)	0.8	(0)	130	(0)	12
14	(0)	10	(0)	0.8	(0)	130	(0)	12
15	(0)	10	(0)	0.8	(0)	130	(0)	12
16	(1)	15	(0)	0.8	(−1)	90	(0)	12
17	(0)	10	(0)	0.8	(1)	170	(−1)	9
18	(0)	10	(1)	1.0	(−1)	90	(0)	12
19	(−1)	5	(0)	0.8	(−1)	90	(0)	12
20	(0)	10	(0)	0.8	(0)	130	(0)	12
21	(0)	10	(−1)	0.6	(−1)	90	(0)	12
22	(0)	10	(−1)	0.6	(1)	170	(0)	12
23	(0)	10	(0)	0.8	(−1)	90	(1)	15
24	(0)	10	(1)	1.0	(0)	130	(1)	15
25	(−1)	5	(−1)	0.6	(0)	130	(0)	12
26	(0)	10	(−1)	0.6	(0)	130	(−1)	9
27	(0)	10	(1)	1.0	(1)	170	(0)	12
28	(−1)	5	(0)	0.8	(0)	130	(1)	15
29	(0)	10	(−1)	0.6	(0)	130	(1)	15

3.3 Evaluation of Gear Quality and WSEM Productivity

For better operating performance and service life, the manufacturing processes of gears should impart geometric accuracy as well as good surface quality and that too with productivity. Consequently, micro-geometry, macro-geometry and surface roughness were selected to evaluate the quality of miniature gears and volumetric gear cutting rate (GCR) was selected to evaluate the productivity of WSEM process. Measurement and evaluation procedure for these parameters are described in the following paragraphs.

3.3.1 Evaluation of Micro-geometry and Macro-geometry of Miniature Gears

Micro-geometry, macro-geometry and flank surface topography of miniature gears were measured on the computer numerically controlled (CNC) gear metrology machine ***SmartGear 500*** from ***Wenzel GearTec***, ***Germany*** (refer Fig. 3.3b) with

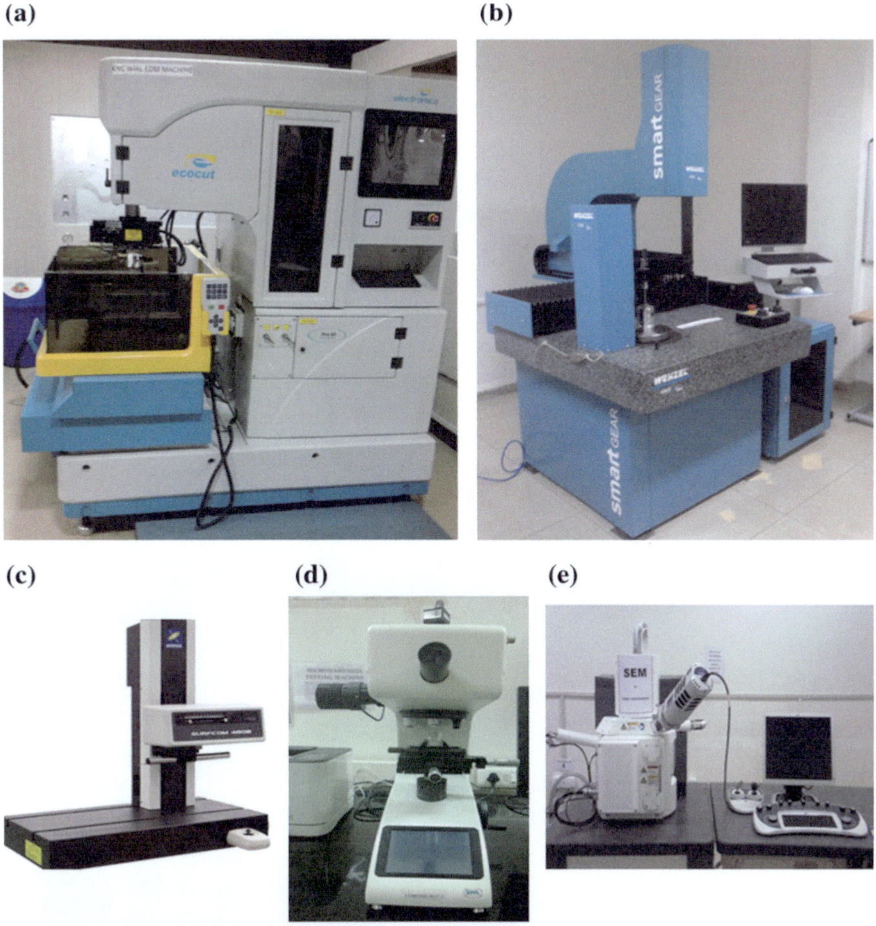

Fig. 3.3 Machine, equipment and instruments used during machining and measurement of miniature gears **a** WSEM machine tool; **b** CNC gear metrology machine; **c** roughness tester; **d** micro-hardness tester; **e** scanned electron microscope

the help of 0.5 mm diameter probe. The maximum permissible error (MPE) of the gear metrology machine is 5 μm. Figure 3.4a–c depicts the measurement procedure for profile and pitch errors.

For total profile error, the measurements at least over four teeth are recommended; therefore, the left and right flanks of four equally spaced gear teeth (namely teeth no. 1; 4; 7 and 10) were traced by the probe along the functional

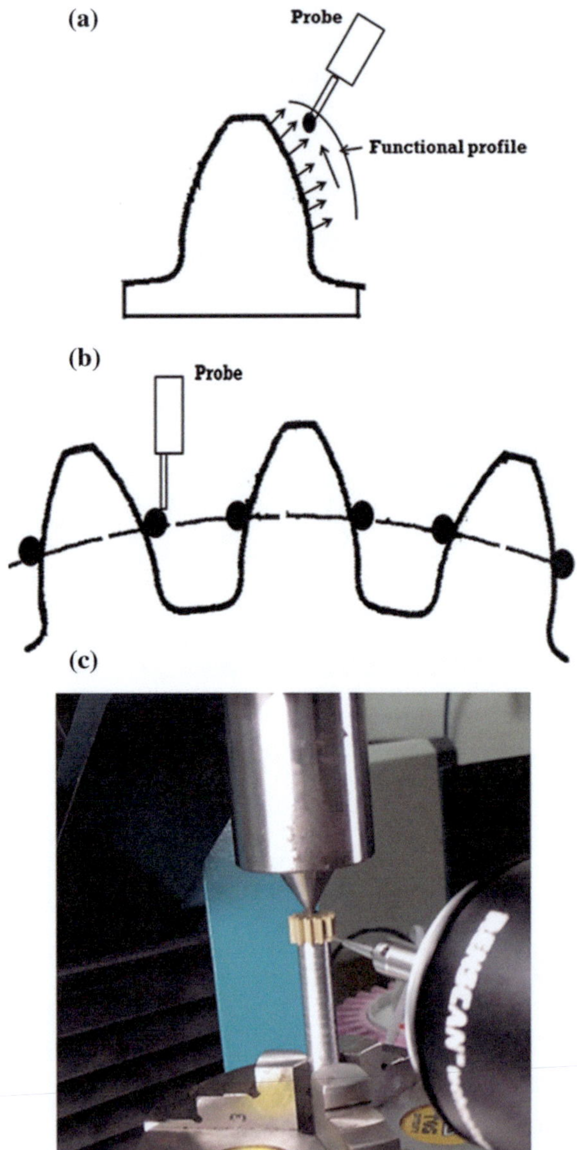

Fig. 3.4 Procedure of measurement of **a** profile error, **b** pitch error of an external gear; and **c** measuring probe of metrology machine tracing the gear [5]

profile (as shown in Fig. 3.4a) on the gear metrology machine (Fig. 3.4c). For each gear, the F_a was calculated by taking average of the mean values of the profile errors in four left flanks (LF) and four right flanks (RF). For measurement of accumulated pitch error, the probe touches the left and right flank surfaces of all gear teeth at a point located at the centre of the face width and on the pitch line (Fig. 3.4b). In this way machine measures the actual pitches for all twelve flanks of the same side. For a particular gear, the F_p values of twelve right flanks and of twelve left flanks were averaged out. DIN-3962 standard was used for assigning the quality to the gears. This standard contains the tolerances for cylindrical gears of various modules, pitch circle diameter and face width. The measured values of F_a and F_p were compared with the tolerance specified for them in the DIN-3962 standard and on the basis of allowable deviation for a particular quality class the gears are assigned certain numbers.

Gear metrology machine also measures macro-geometry parameters such as span, chordal tooth thickness and outside diameter (i.e. diameter over-the-two-balls). These parameters were measured only for the best quality gear manufactured during the pilot experiments and at the optimum parameters from the multi–objective optimization. Based on the specifications of the miniature gears, the span was measured for two teeth. Ball diameter of 1.236 mm was found suitable for measurement of dimensions over-the-two-balls. Chordal tooth thickness was measured by touching the probe at both the sides of the tooth of a gear. Deviations in these parameters from their theoretical values were used as performance measures.

For the measurement of flank surface topography, the whole surface of both the flanks (i.e. right flank and left flank) of the tooth of best miniature gear divided into nine equal segments along face width and four equal segments along tip to root. These segments are represented in the form of a grid. Grid made of blue straight lines is used to represent the theoretical topography, whereas the grid made of curved black color lines for representing the actual topography of tooth flank surface.

3.3.2 Evaluation of Surface Roughness

The evaluation of surface roughness for trial experiments was done using ***Perthometer*** of ***Mahr*** **Germany** by taking measurements at three different locations on brass strips keeping evaluation and cut-off length as 0.75 and 0.25 mm respectively. For pilot, main and confirmation experiments the surface roughness parameters (average roughness, maximum roughness, bearing area parameters, skewness and kurtosis) were evaluated using ***Surfcom*** roughness profiler from ***Accretech***, ***Japan*** (refer Fig. 3.3c) on an evaluation length of 0.75 mm on gear tooth flank surface along root to tip (across the wire feeding direction) using 0.25 mm as the cut-off length. The '*R-curve*' option was used as form removal i.e. for compensating the curvature effect of the gear tooth geometry during roughness inspection. The roughness parameters were filtered-out by *Gaussian* filter. The measurements were taken at three different places on a single tooth of a gear for each experimental run (having two replicates) and average of the six values was considered.

3.3.3 *Characterization of Subsurface Integrity*

3.3.3.1 Microstructure Examination

The physical nature of the surface of the miniature gear tooth was evaluated by analysis of the images generated on scanning electron microscope ***SUPRA 55*** from ***Carl Zeiss Germany*** (refer Fig. 3.3e) at various levels of magnification. The samples were polished in three stages using SiC emery paper of three different grit sizes of 200, 500 and 800 sequentially. The polished samples were mild etched by swabbing using ferric chloride solution prior to their microstructure examination.

3.3.3.2 Micro-hardness Measurement

To find the presence and depth of the recast layer the *Vickers* micro-hardness of the miniature gear tooth surface was measured on micro-hardness testing machine model from **VMH002 V** from ***Walter UHL Technische Mikroskopie GmbH***, ***Germany*** (refer Fig. 3.3d). The readings were taken at different indentation depths by varying the test load in the range of 15–200 g. The indentation speed and dwell time were kept at 25 μm/s and 15 s respectively during the testing. Three replicates were taken and average micro-hardness value was used for analysis.

3.3.4 *WSEM Process Productivity and Evaluation*

For any manufacturing process, productivity is a major concern for its commercial adoption along with its capability to produce components having high quality and high accuracy. In the present work, the productivity of the WSEM process was studied in terms of volumetric gear cutting rate (GCR). Average value of the volumetric GCR was obtained by dividing the mass loss of the gear blank during WSEM of gears by the product of machining time and density of the gear material (i.e. brass) as represented by Eq. (3.3). The mass of the gear blank was measured on a precision weighing balance (make *Essae-Teraoka Ltd.*) having a least count of 0.01 g.

$$Gear\ Cutting\ Rate\ (GCR) = \frac{Mass\ of\ gear\ blank\ before\ WEDM\ (\mathrm{gm}) - Mass\ of\ gear\ blank\ after\ WEDM\ (\mathrm{gm})}{Machining\ time\ (\mathrm{min}) \times Density\ of\ gear\ material\ (\mathrm{gm/mm^3})}\ (\mathrm{mm^3/min}) \tag{3.3}$$

3.4 Procedure of Experimentation

Ecocut CNC wire-EDM/SEM machine (see Fig. 3.3a) from Electronica Machine Tools Pvt. Ltd. Pune was used for manufacturing the miniature spur gears. This machine is based on closed-loop control system and having tolerance of ±15 μm. Figure 3.5 depicts the sequence of manufacturing the miniature gears by WSEM. Rectangular brass plate of 5 mm thickness was used as raw material i.e. gear blank. Part programs for manufacturing the gears on the WSEM machine were prepared by *Elcam* software offline in a computer attached to the WSEM machine. The *Elcam* software has a separate subroutine for generating CNC part program for the defined gear specifications. The compensation (off-set) for wire diameter and machining overcuts can also be specified through this software. After feeding the required input the software automatically generates all the numerical information about the movement of the worktable in terms of geometric and miscellaneous codes (G and M codes). The entered gear geometry was verified on the graphic display screen using the simulation facility. The CNC part program was then transferred to the WSEM machine by RS 232 cable (refer Appendix B for part program to cut the miniature gear of given specification by WSEM using the wire of 250 μm diameter).

The gears manufactured by WSEM machine were tested on gear metrology machine, roughness tester and on other applicable instruments. Optimization techniques were then used to optimize the WSEM parameters for further improving the quality and increasing the productivity of miniature gears. Experiments were

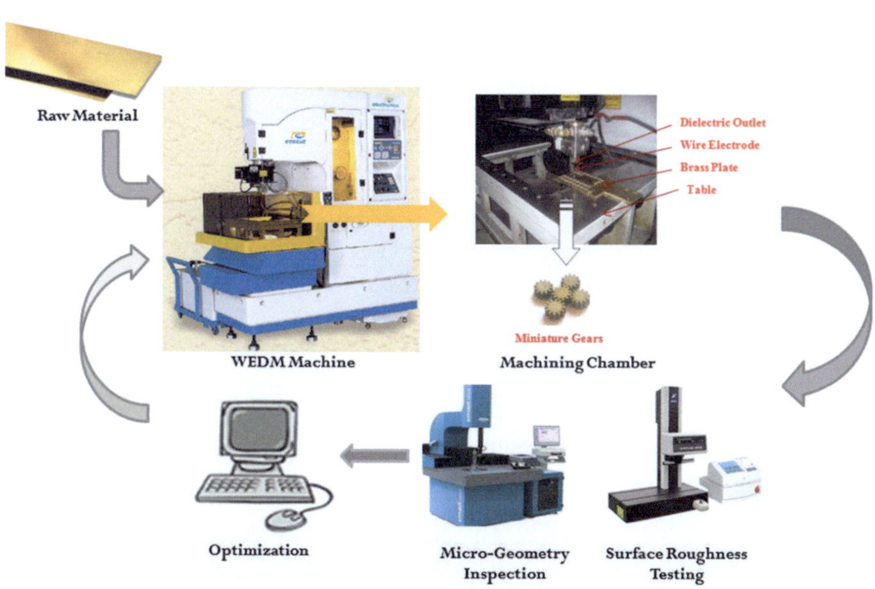

Fig. 3.5 Tasks performed for meso-fabrication of miniature gears by WSEM [6]

conducted at the optimized parameters and manufactured gears were inspected/ tested again to validate the results of optimization.

3.5 Methods and Tools for Data Analysis

The results of trial experiments were analyzed by studying the effects of variable WSEM parameters on the wire breakage frequency, surface roughness and the cutting speed. The results of pilot experiments were analyzed by studying the variation of the considered responses with each variable WSEM process parameter. The results of the main experiments were used to develop response surface models. The adequacy of these models was checked using the different statistical tools. Subsequently, results of main experiments were analyzed through analysis of variance (ANOVA) to determine the relative importance of the input parameters and their interactions. The details of the tools and methods used in the analysis of the experimental data are explained in the following sections.

3.5.1 Analysis of Variance (ANOVA)

Analysis of variance (ANOVA) is the statistical technique used to analyze the experimental data which helps in determining the significance of the process parameters and the interaction effects on the process responses. The ANOVA also gives an estimate of the error variance for the main effects and variance of the prediction error. The total variability in the experimental observations is partitioned into two components namely (i) the variability among the different experiments; and (ii) variability within a particular experiment.

ANOVA involves following steps:

1. **Total sum of squares (SST)**

$$SST = \sum_{i=1}^{n}\sum_{j=1}^{n_r} y_{ij}^2 - \frac{1}{nn_r}\left(\sum_{i=1}^{n}\sum_{j=1}^{n_r} y_{ij}\right)^2 \quad (3.4)$$

Here, y_{ij} is the value of response for jth *replication of the* ith *experiment*; *n is the total number of experiments*; *and n_r is the number of replications for each experiment.*

2. **Sum of squares (SS) for the main effects**: The sum of square for any parameter p for all its level is given by

$$(SS_{main})_p = \sum_{l=1}^{p_l}\frac{1}{n_{p_{lt}}}\sum_{k=1}^{n_{p_{lt}}}\left(\frac{1}{n_r}\sum_{j=1}^{n_r} y_{lkj}\right)^2 - \frac{1}{nn_r}\left(\sum_{i=1}^{n}\sum_{j=1}^{n_r} y_{ij}\right)^2 \quad (3.5)$$

Here, y_{lkj} *is the value of the reponse for the* jth *replication of* kth *experimental observation at* lth *level of the* pth *parameter;* p_l *is the number of levels for* pth *parameter and* $n_{p_{lt}}$ *is the total number of experimental observations for* lth *level of the* pth *parameter.*

3. **Sum of squares (SS) for interaction effects**: Since the interaction effects are in two columns, therefore the sum of squares for any *i*th interaction effect $(SS_{interaction})_i$ has been calculated by adding the sum of square for each interaction column as follows

$$(SS_{interaction})_i = SS_{interaction\ col.1} + SS_{interaction\ col2} \quad (3.6)$$

4. **Sum of squares (SS) for errors**:

$$SS_{errors} = SST - \sum_{p=1}^{n_p} (SS_{main})_p - \sum_{i=1}^{n_i} (SS_{interaction})_i \quad (3.7)$$

Here, n_p *is the total number parameters and* n_i *is the total number of interactions.*

5. **Mean sum of squares (MSS)**: Mean sum of square for *p*th parameter is given by

$$MSS_p = \frac{(SS_{main})_p}{v_p} \quad (3.8)$$

Here, MSS_p *is the Mean sum of squares due to parameter p; and* v_p *is number of degress of freedom associated with it.*

6. **Mean sum of squares (MSS) for errors**: Mean sum of square for the errors MSS_{errors} is given by

$$MSS_{errors} = \frac{SS_{errors}}{v_{errors}} \quad (3.9)$$

Here, v_{errors} *is number of degress of freedom associated with the errors.*

7. ***F*-ratio**: Fisher's ratio for p parameter is calculated using

$$F_p = \frac{MSS_p}{MSS_{errors}} \quad (3.10)$$

Here, F_p *is the value of the F-ratio for the parameter p; and* MSS_e *is the value of the F-ratio for the error.*

8. **Comparison of *F*-ratios**: The value of *F-ratio* calculated for each parameter is compared with the value available in the standard tables corresponding to the DOF of that parameter for different confidence levels. If the calculated value of the *F-ratio* is greater than the tabulated value of the *F-ratio* at any confidence level, then the parameter is significant at that confidence level otherwise it is insignificant.

3.5.2 Model Adequacy Checking

The relationship between variables and responses of any process or system are developed by the regression models coupled with response surface designs. Before the conclusions from the ANOVA are adopted, the adequacy of the underlying model should be checked. Model adequacy checking is an important part of the data analysis procedure. It is always necessary to examine the fitted model to ensure that it provides an adequate approximation to the true system and to verify that none of the least squares regression assumptions are violated. The regression model will probably give poor or misleading results unless it is an adequate fit. Following parameters were used to examine the adequacy of the model:

1. ***Residual analysis***: Violation of basic assumptions and model adequacy can be easily investigated by the examination of residuals. If the model is adequate, the residuals should be structure-less, i.e. they should contain no obvious pattern. Two important plots namely normal probability plot and plot of residuals versus fitted values are the tools in the form of graphical analysis for model diagnostic checking. If the underlying error distribution is normal then normal probability plot will resemble a straight line. It is plotted for *studentized* (residual divided by its estimated standard deviation) residuals. No obvious pattern of the residual versus fitted value graph shows the adequacy of the model adopted.

2. ***Lack of fit test***: It compares the residual error with the pure error from the replicated design points. The insignificant values of *lack of fit* i.e. p value >0.05 of the model confirms the adequacy of the adopted model.

3. ***R-squared statistic***: It is a measure of goodness of fit. It is expressed by

$$R^2\ statistics = \frac{variability\ explained\ by\ the\ model}{total\ variability\ in\ the\ actual\ data} \tag{3.11}$$

The more it approaches unity, the better model fits the experimental data.

Predicted R-squared measures the amount of variation in new data explained by the model. Its value close to 1 is preferred.

Adjusted R-Squared is the R-squared value after adjusting for the number of terms in the model relative to the number of design points.

The values of *predicted R-Squared* and *adjusted R-Squared* should be in good agreement to have a high correlation between the observed and predicted values.

4. **PRESS value**: *Prediction error sum of squares* (*PRESS*) is a measure of how well the model fits each point in the design. In other words it is a measure of how well the model for the experiment is likely to predict the responses in a new experiment. Smaller the PRESS statistics, better the model fits the data. It is advisable to choose the model which is having lowest PRESS value.

5. ***Adequate precision***: Adequate precision measures the signal to noise ratio and is computed using

$$Adequate\,precision = \frac{maximum\,predicted\,response - minimum\,predicted\,response}{Avg.\,standard\,deviation\,of\,all\,predicted\,responses} \tag{3.12}$$

Adequate precision value >4 is desirable.

References

1. Ross PJ (1988) Taguchi techniques for quality engineering. Tata McGraw-Hill, New York
2. Mitra A (2010) Fundamentals of quality control and improvement. Prentice Hall of India, New Delhi
3. Phadke MS (1989) Quality engineering using robust design. Prentice-Hall Inc., New Jersey
4. Montgomery DC (2001) Design and analysis of experiments. John-Wiley & Sons Inc., New York
5. Gupta K, Jain NK (2014) Analysis and optimization of micro-geometry of miniature gears manufactured by wire electric discharge machining. Precis Eng 38(4):728–737
6. Gupta K, Jain NK (2014) Comparative study of wire-EDM and hobbing for manufacturing high quality miniature gears. Mater Manuf Processes 29:1470–1476

Chapter 4
Experimental Results and Analysis

4.1 Conclusions from Trial Experiments

Figure 4.1 depicts the input to and outcome from each experimental stage clarifying the role and importance of each stage.

The bracketed ranges of WSEM parameters after conducting forty trial experiments were 5–25 V for voltage, 0.6–1.4 μs for pulse-on time, 90–250 μs for pulse-off time, and 3–15 m/min for wire feed rate. Whereas, the available ranges on the WSEM machine are 1–100 V for voltage; 0.1–1.65 μs for pulse-on time; 10–260 μs for pulse-off time; and 1–15 m/min for wire feed rate. These ranges were bracketed on the basis of the following observations [1]:

- Voltage less than 5 V and pulse-on time less than 0.6 μs were resulted in unstable machining at very slow cutting rate. Whereas, machining at voltage between 25 and 45 V caused of deterioration of the surface finish. The spark gap seems to increase while using voltage more than 45 V thus leading to unstable machining at slow cutting rate.
- High frequency of wire breakage was observed at pulse-on time more than 1.4 μs and at pulse-off time less than 90 μs.
- Machining below 3 m/min wire feed rate led to high frequency of wire breakage and high surface roughness.

4.2 Results of Pilot Experiments and Their Analysis

Pilot experiments were designed and conducted based on one factor at-a-time approach. The ranges of WSEM parameters were bracketed by analyzing the trend of the variation of responses i.e., micro-geometry parameters and surface roughness

© Springer Science+Business Media Singapore 2017
K. Gupta and N.K. Jain, *Near-Net Shape Manufacturing of Miniature Spur Gears by Wire Spark Erosion Machining*, Materials Forming, Machining and Tribology, DOI 10.1007/978-981-10-1563-2_4

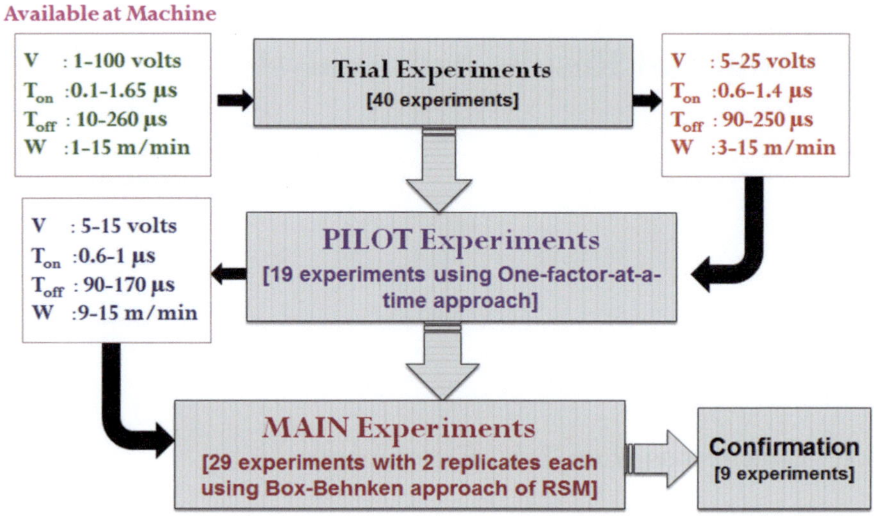

Fig. 4.1 Flow-chart showing different experimental stages with corresponding inputs and outcomes

parameters with them. The detailed analysis of the responses is described in the following sections.

4.2.1 Analysis of Micro-geometry

Table 4.1 presents values of micro-geometry parameters i.e., total profile error 'F_a' and accumulated pitch error 'F_p' for different runs of pilot experiments. It is evident from this table that the experimental combination of 15 V as voltage; 1.0 µs as T_{on}; 170 µs as T_{off}; 9 m/min as wire feed rate and full cutting speed (at experiment no. 23) manufactured the gear having minimum values of 'F_a' and 'F_p', which then recognized as the the best quality gear.

Figure 4.2 depicts the variation of the total profile error 'F_a' and accumulated pitch error 'F_p' with four WSEM process parameters in terms of experimental points and the best fit 2nd order polynomial curves. It is evident that the 'F_a' is increasing with increase in voltage (Fig. 4.2a), pulse-on time (Fig. 4.2b), and pulse-off time (Fig. 4.2c) while, it is decreasing with increase in wire feed rate (Fig. 4.2d). Since, gear performance, particularly noise, is very sensitive to total profile error; therefore, even a slight change in it significantly affects the performance characteristics of the gears. The variation of the accumulated pitch error 'F_p' confirms the existence of an optimum range of voltage (Fig. 4.2a), pulse-on time (Fig. 4.2b), and pulse-off time (Fig. 4.2c) and selection of higher value of wire feed rate due to its decreasing trend with this.

Table 4.1 Values of variable input parameters with corresponding values of 'F_a' and 'F_p' for different runs of pilot experiments

Expt. No.	Variable input parameters					Responses					
	V (V)	T_{on} (μs)	T_{off} (μs)	W (m/min)	F (%)	Total profile error 'F_a' (μm)			Accumulated pitch error 'F_p' (μm)		
						Left Flank	Right Flank	Average	Left Flank	Right Flank	Average
1	05	1.0	170	9	75	13.70	12.80	13.25	27.40	29.40	28.40
2	10				75	15.90	11.30	13.60	21.80	23.60	22.70
3	15				75	15.80	12.70	14.25	22.50	23.70	23.10
4	20				75	15.70	11.70	14.35	24.30	24.50	24.40
5	25				75	15.90	17.00	16.45	27.60	22.10	24.85
6	15	0.6	170	9	75	12.40	11.20	11.80	37.80	27.30	32.55
7		0.8			75	15.40	11.40	13.40	16.60	17.10	16.85
8		1.0			75	15.80	12.70	14.25	22.50	23.70	23.10
9		1.2			75	16.70	15.90	16.30	39.90	45.10	42.50
10		1.4			75	14.60	18.90	16.75	52.40	53.40	52.90
11	15	1.0	90	9	75	14.00	11.10	12.55	12.70	16.60	14.65
12			130		75	15.40	12.10	13.75	21.80	21.10	21.45
13			170		75	15.80	12.70	14.25	22.50	23.70	23.10
14			210		75	15.70	12.90	14.30	37.80	40.40	39.10
15			250		75	18.10	12.40	15.25	67.20	47.30	57.25
16	15	1.0	170	3	75	13.20	15.60	14.40	39.80	43.40	41.60
17				6	75	15.70	12.70	14.30	37.80	40.40	39.10
18				9	75	15.80	12.70	14.25	22.50	23.70	23.10
19				12	75	14.70	11.30	13.70	25.90	19.20	22.55
20				15	75	16.20	11.20	13.00	17.10	15.30	16.20
21	**15**	**1.0**	**170**	**9**	50	12.50	16.30	14.40	29.70	27.50	28.60
22					75	15.80	12.70	14.25	22.50	23.70	23.10
23					**100**	**13.30**	**13.10**	**13.20**	**9.70**	**12.70**	**11.20**

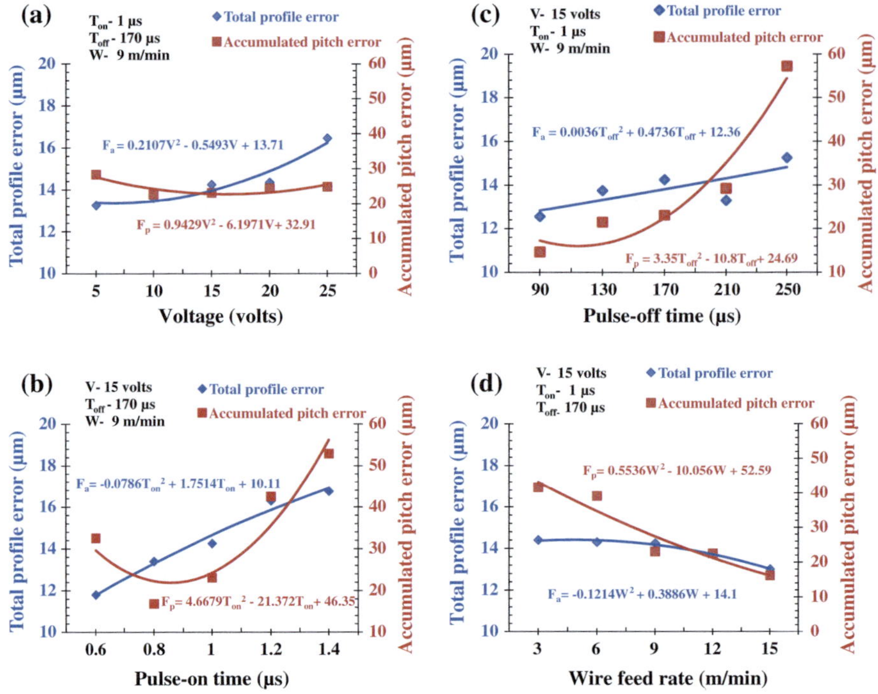

Fig. 4.2 Variation of total profile error 'F_a' and accumulated pitch error 'F_p' with **a** voltage; **b** pulse-on time; **c** pulse-off time; and **d** wire feed rate [3]

The main reasons responsible for errors in micro-geometry of WSEMed products are (i) generation of irregular shaped craters by the violent sparks produced by high energy discharge parameter settings, and (ii) *wire-lag* i.e., deviation of wire from its path due to various forces acting on the wire such as mechanical forces produced by the pressure of the gas bubbles, axial forces applied to straighten the wire, the hydraulic forces induced by the flushing, electro-dynamic forces inherent to the spark generation and the electro-static forces [2–7]. Increments in micro-geometry errors i.e., F_a and F_p of WSEMed miniature gears with increase in voltage, pulse-on time and pulse-off time, and with decrease in wire feed rate can be better explained by the following facts:

- Increase in discharge energy with higher voltage (and in turn peak current) and longer pulse-on time causing the formation of deeper and irregular craters, and increase in thermal stresses which in turn produces rippled surface with change in the microstructure [4, 8].
- Increase in electro-dynamic forces due to high spark at higher voltage and longer pulse-on time deviate wire from its path and thereby increase in F_a and F_p [3–7].

- Increase in material removed in a single discharge cycle with increase in pulse-on time causing wire-lag due to the generation of excessive forces produced by pressure of the gas bubbles [3–7].
- High pulse-off time increases the duration of dielectric flow, which is more than the required flushing time. This can generate excessive hydraulic force, which leads to wire-lag [2, 3].
- High wire feed rate resulted in low errors because of the availability of very short period for spark concentration therefore, smaller sized and flat craters were believed to have been generated.

Consequently, it was concluded that for minimizing the total profile error 'F_a' and accumulated pitch error 'F_p' simultaneously, optimum ranges of voltage (5–15 V), pulse-on time (0.6–1 μs) and pulse-off time (90–170 μs); and higher values of wire feed rate and cutting speed should be selected in the WSEM of miniature gears [3].

4.2.2 *Analysis of Surface Roughness*

Table 4.2 presents the values of the surface roughness parameters i.e., average surface roughness 'R_a' and maximum surface roughness 'R_{max}' for different runs of pilot experiments.

Effects of voltage, pulse-on time, pulse-off time, and wire feed rate on 'R_a' and 'R_{max}' are depicted by Fig. 4.3a–d in terms of the experimental points and the best fit 2nd order polynomial curves. Some of the graphs are not passing through all the experimental points. It is evident from Fig. 4.3a that R_a increases almost linearly with the voltage, while variation trend of R_{max} is non-linear with a minimum value occurring in the range of 5–10 V. Variation trend of both R_a and R_{max} with pulse-on time as shown in the Fig. 4.3b is non-linear with minimum R_a occurring at 0.6 μs and minimum R_{max} occurring in the range of 0.6 to 0.8 μs. In other words, the use of higher voltage and longer pulse-on time leads to generation of comparatively rough surface mainly due to formation of the deeper and irregular craters by violent sparks which occur at higher discharge energy parameter settings [8–10].

Figure 4.3c shows that there is a non-linear variations of R_a and R_{max} with pulse-off time; the minimum values of R_a exist in the range of 210–250 μs, while minimum R_{max} occurring at 250 μs pulse-off time. Decrease in the pulse-off time for a constant value of pulse-on time is responsible for high discharge. This consequently results in high surface roughness due to the poor flushing of the molten material from the machining zone. Moreover, short-circuiting caused by the residual or un-expelled molten material in the machining zone also increases the surface roughness. As depicted in Fig. 4.3d, the R_{max} decreases almost linearly with the wire feed rate, whereas the variation of R_a with wire feed rate is non-linear with minimum value (0.96 μm) occurring at 15 m/min. The reason behind the decreasing trend of surface roughness with wire feed rate is the fact that higher wire

Table 4.2 Values of variable input parameters with corresponding values of 'R_a' and 'R_{max}' for different runs of pilot experiments

Expt. No	Input parameters					Responses	
	V (V)	T_{on} (μs)	T_{off} (μs)	W (m/min)	F (%)	Average roughness 'R_a' (μm)	Maximum roughness 'R_{max}' (μm)
1	5	1	170	9	75	1.06	7.46
2	10					1.57	7.40
3	15					1.46	7.84
4	20					1.80	10.3
5	25					2.17	13.3
6	15	0.6	170	9	75	1.03	7.28
7		0.8				1.22	6.62
8		1				1.46	7.84
9		1.2				1.43	7.94
10		1.4				1.55	9.42
11	15	1	90	9	75	1.60	8.0
12			130			1.22	8.40
13			170			1.46	7.84
14			210			1.29	7.56
15			250			1.27	6.78
16	15	1	170	3	75	1.46	9.62
17				6		1.43	7.12
18				9		1.46	7.84
19				12		1.13	6.44
20				15		0.96	4.90
21	15	1	170	9	50	1.48	8.4
22					75	1.46	7.84
23					100	1.0	6.4

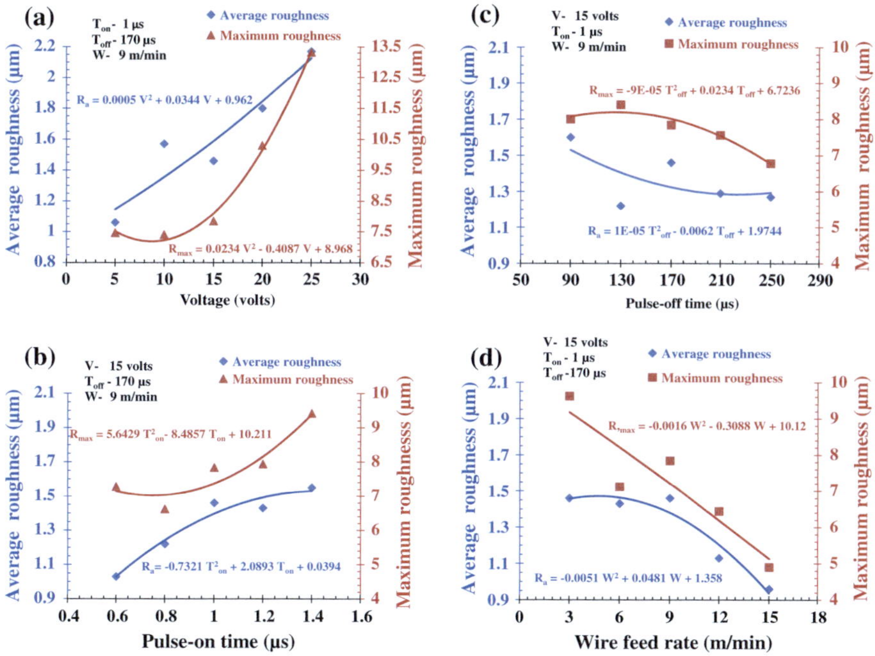

Fig. 4.3 Variation of average roughness 'R_a' and maximum roughness 'R_{max}' with WSEM process parameters **a** voltage; **b** pulse-on time; **c** pulse-off time; and **d** wire feed rate

feed rate decreases the frequency of wire breakage and provides less time to the wire to adhere on flank surface of gear tooth consequently, increasing the surface finish [9, 10]. Moreover, both R_a and R_{max} were found to be decreased with increase in the cutting speed as stable machining takes place at higher cutting speed which consequently improves the surface finish.

Keeping in view the above mentioned observations, an objective of simultaneously minimizing total profile error, accumulated pitch error, average and maximum roughness, and giving more importance to micro-geometry errors than surface roughness, the ranges of WSEM parameters bracketed for the further investigations are 5–15 V for voltage; 0.6–1.0 µs for pulse-on time; 90–170 µs for pulse-off time; 9–15 m/min for wire feed rate and maximum cutting speed.

4.2.3 Analysis of the Best Quality Miniature Gear

The miniature gear manufactured in the pilot experiment number 23 at parametric combination of 15 V as voltage, 1 µs as pulse-on time, 170 µs as pulse-off time, 9 m/min as wire feed rate and at maximum cutting speed had the lowest errors in total profile and accumulated pitch, and consequently it was considered as the 'best quality gear'. All the quality and surface integrity aspects of this gear were investigated in details, which are discussed in the subsequent sections.

4.2.3.1 Micro-geometry

The best quality miniature gear had *DIN* quality number 8 for profile (with total profile error of 13.2 μm) and *DIN* quality number 6 for pitch (with accumulated pitch error of 11.2 μm). The graphs of Fig. 4.4 illustrate the results of F_a and F_p for the best quality gear.

Figure 4.4a depicts the graph of total profile error (F_a) for left flank (LF) and for right flank (RF) of four teeth (namely tooth no. 1; 4; 7 and 10) and their mean (x). Blue color vertical straight lines indicate best fit line and the black color curvatures represent the deviations of profiles i.e., profile errors. The total profile error (F_a) for this gear is 13.2 μm, calculated by taking average of the mean values of the profile error of four left flanks (13.3 μm) and profile error of four right flanks (13.1 μm). The figure also contains the information about profile form (ff_a) and profile slope (fh_a) errors for the same set of teeth. Figures 4.4b and 4.4c depict the accumulated pitch error (F_p) for left flanks and right flanks respectively. The accumulated pitch error (F_p) for this gear is 11.2 μm, calculated by taking average of accumulated pitch error for left flanks (9.7 μm) and accumulated pitch error for right flanks (12.7 μm). Although the lead error was not investigated and analyzed as response parameter, yet it was measured for the best quality gear. The value of the total lead error (F_β) for this gear is 5.4 μm (DIN 7) as illustrated in the Fig. 4.5.

4.2.3.2 Flank Surface Topography

Figure 4.6 illustrates the topography of the left and right flanks of one of the teeth of the best gear.

The entire surface of both the flanks (i.e., right flank and left flank) of the tooth of best miniature gear divided into nine equal segments along face width and four equal segments along tip to root. Grid formed by these segments enhances the analysis of flank surface at any particular region. The grid made of blue straight lines represents the theoretical topography whereas the grid made of curved black color lines represents the actual topography of tooth flank surface. It is evident from the figure that the actual topography of a tooth of the best gear indicates very small deviation from the theoretical one.

4.2.3.3 Macro-geometry

Very low values of macro-geometry errors i.e., deviation in span (4 μm), deviation in chordal tooth thickness (5 μm) and deviation in the outside diameter (10 μm) were measured for the best quality miniature gear.

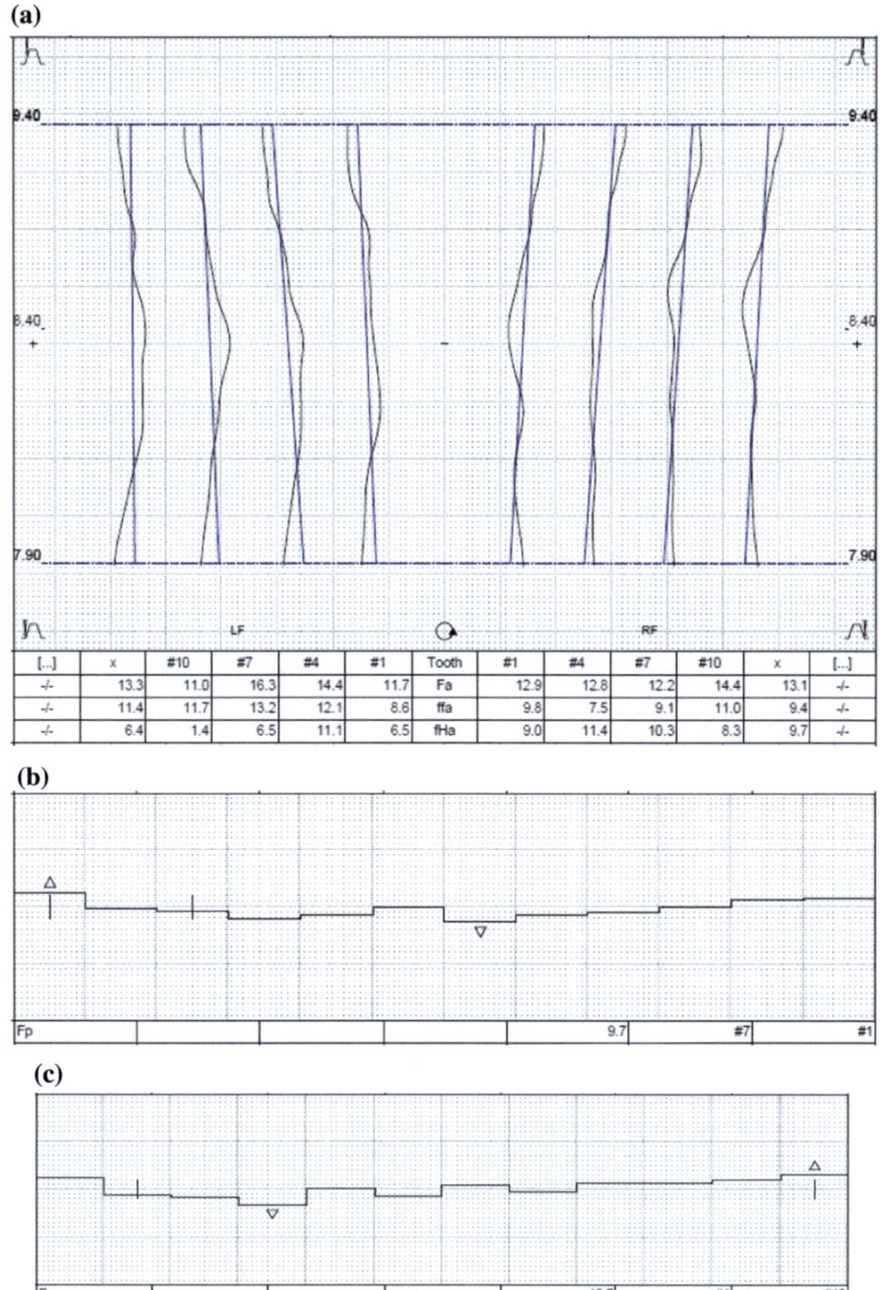

[...]	x	#10	#7	#4	#1	Tooth	#1	#4	#7	#10	x	[...]
-/-	13.3	11.0	16.3	14.4	11.7	Fa	12.9	12.8	12.2	14.4	13.1	-/-
-/-	11.4	11.7	13.2	12.1	8.6	ffa	9.8	7.5	9.1	11.0	9.4	-/-
-/-	6.4	1.4	6.5	11.1	6.5	fHa	9.0	11.4	10.3	8.3	9.7	-/-

Fig. 4.4 Micro-geometry errors measurement using CNC gear metrology machine for the best quality miniature gear manufactured in pilot experimentation **a** total profile error; **b** accumulated pitch error for left flank; and **c** accumulated pitch error for right flank

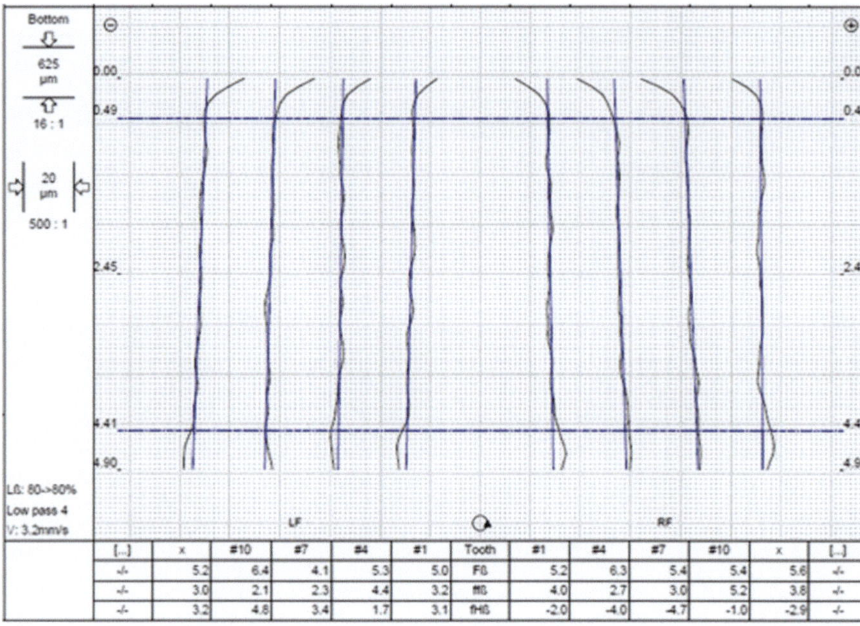

Fig. 4.5 Total lead error for the best quality miniature gear manufactured in pilot experimentation

4.2.3.4 Surface Roughness and Bearing Area Curve

The average surface roughness 'R_a' and maximum surface roughness 'R_{max}' evaluated for this gear were 1 µm and 6.4 µm respectively.

Figure 4.7 presents the bearing area curve for this gear. The values of R_{pk} and M_{r1} are 2.1 µm and 10 % which indicates that about 10 % of initial surface profile area is likely to wear out during initial run-in period with surface peaks reduced by of 2.1 µm depth. The value of R_k is 2.5 µm and corresponding bearing length is 70 % i.e., more than 70 % of the surface roughness data are confined within just 2.5 µm depth. Material component corresponding to the lower limit of the roughness core 'M_{r2}' is about 90 % for the best quality miniature gear. *Reduced valley depth* 'R_{vk}' is 1.8 µm for the 10 % length which indicates quiet good lubricating oil retaining capacity of the gear tooth surface. It is evident from the BAC of the best quality miniature gear that the gear is capable to function well during its service life. It was also observed that the bearing ratio of WSEMed miniature gears decreases with increase in pulse-on time.

4.2.3.5 Microstructure and Micro-hardness

The microstructural investigation of the best miniature gear was also done in order to find the presence of any surface defect such as burrs, cracks, voids, globules and

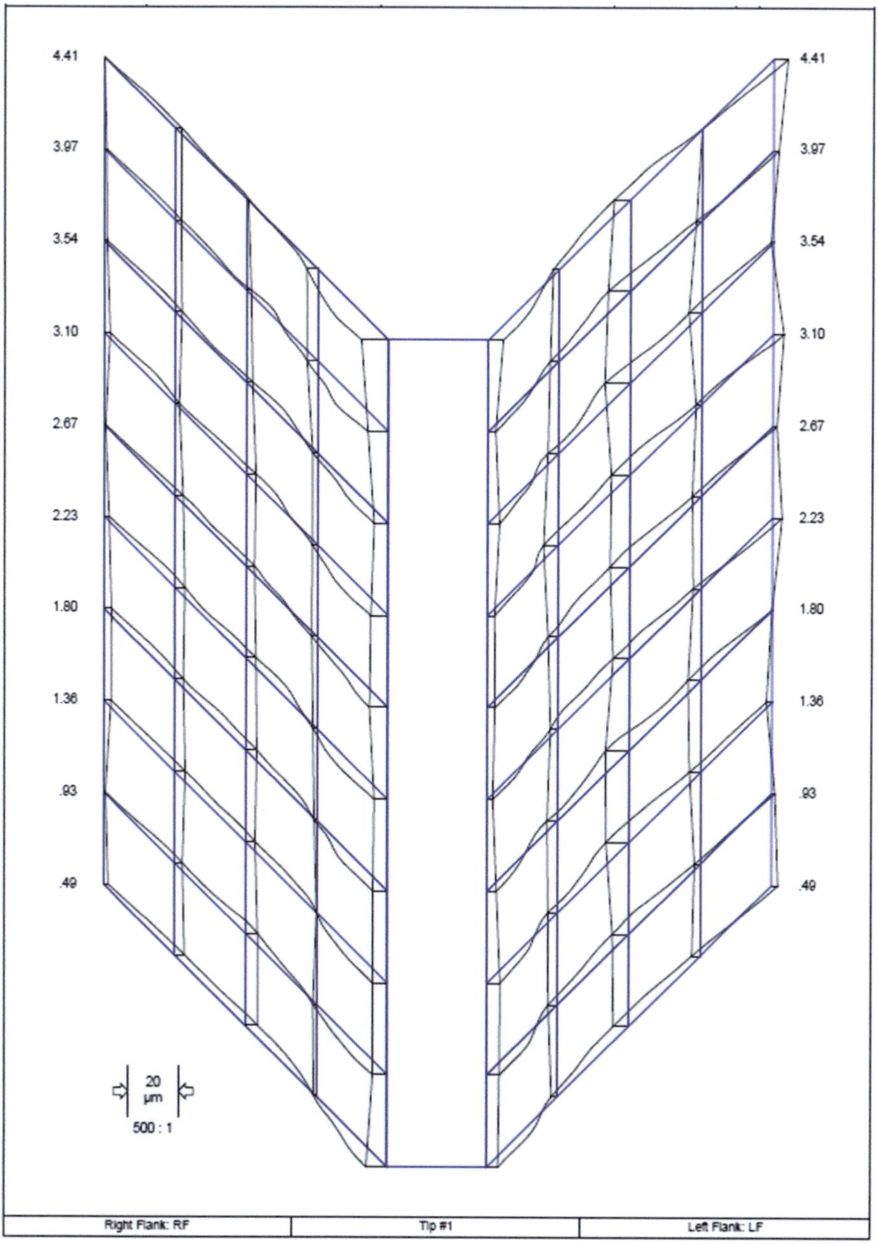

Fig. 4.6 Topography of a tooth of best quality miniature gear manufactured by WSEM during pilot experiments

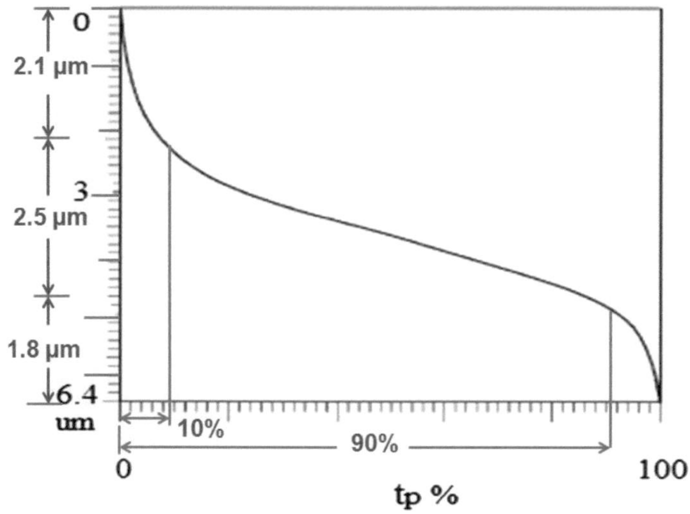

Fig. 4.7 Bearing area curve for the best quality miniature gear

recast layer on the flank surface of the gear tooth. Scanned electron microscope (SEM) images depicted in Fig. 4.8a–d illustrate the uniform teeth profile of the best gear. Whereas, Fig. 4.9a, b present the burr-free surfaces of the teeth of this gear. It

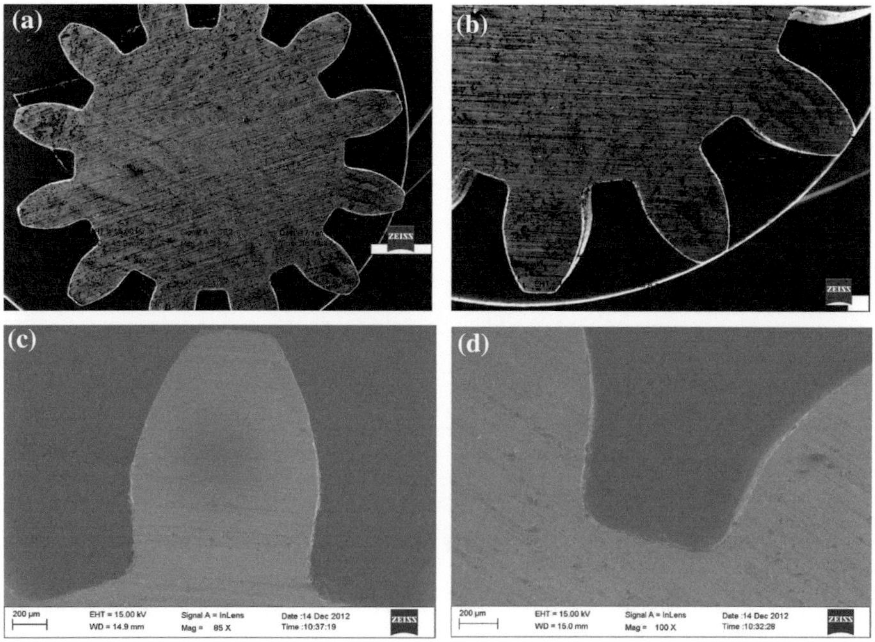

Fig. 4.8 SEM images showing uniform tooth profiles of the best gear. **a** at 20 X, **b** at 70 X **c** at 85 X **d** at 100 X

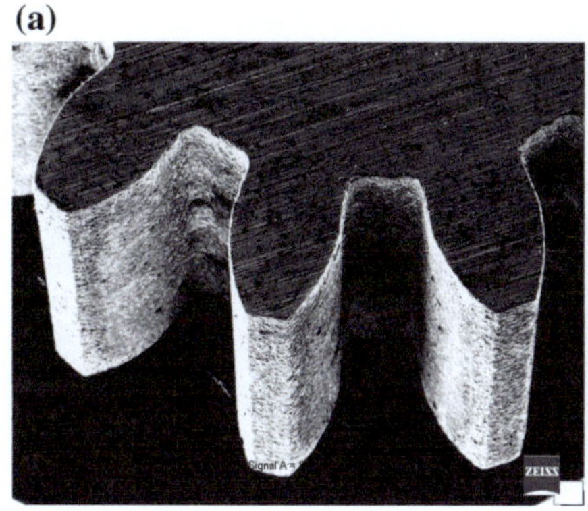

Fig. 4.9 SEM images showing burr-free tooth surfaces of the best gear. **a** at 40 X, **b** at 80 X

was analyzed that the parametric combination 15 V, 1 μs T_{on}, 170 μs T_{off}, and 9 m/min W at which the best quality miniature gear was manufactured kept the discharge energy under control. Crack-free surface with uniform arrangement of regular shaped crater was believed to be generated by rapid dissipation of heat due to high thermal conductivity of brass. Figure 4.10a, b depicts the smooth and surface defect-free microstructure of the tooth of the best gear.

The cratered surface of WSEMed gear tooth is shown by Fig. 4.11. It was observed that high discharge energy parameter settings i.e., voltage more than 15 V and/or pulse-on time more than 1 μs, are the main cause of white layer formation and crack generation on gear teeth. Figure 4.12a presents the tooth surface having white layer, pockmarks and voids, generated at high pulse-on duration (1.4 μs) while, Fig. 4.12b depicts the occurrence of crack on the gear tooth, manufactured at high voltage (above 15 V). High thermal stresses induced at high discharge energy parameter settings were the main cause of formation of white layer and cracks. Selection of an appropriate set of WSEM parameters that would minimize thickness

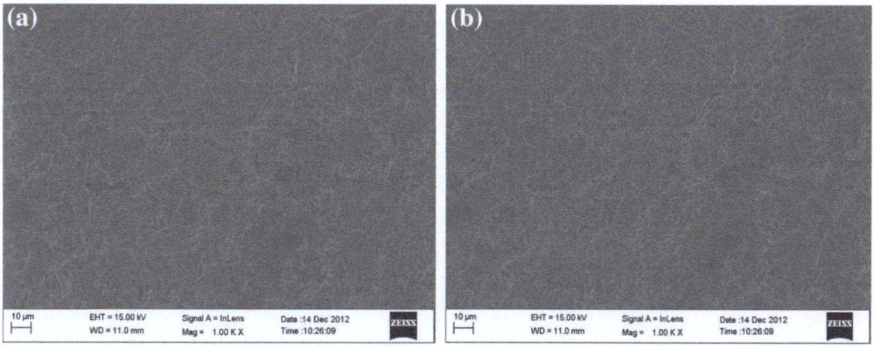

Fig. 4.10 Defect-free microstructure of the tooth surface of the best gear. **a** at 1000 X, **b** at 1000 X

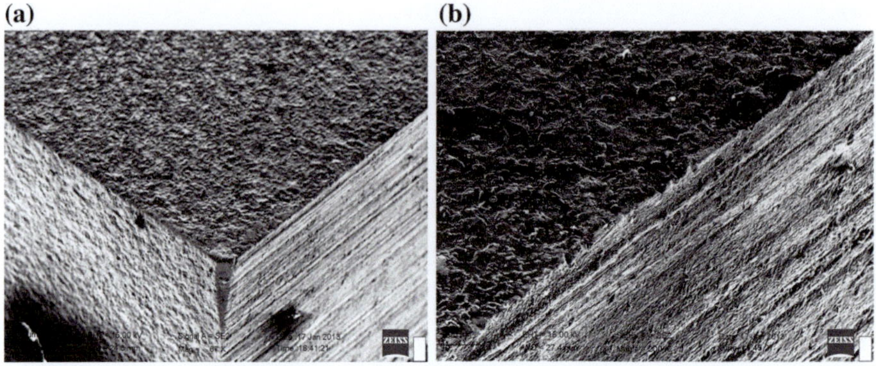

Fig. 4.11 Actual WSEMed surface of the gear tooth. **a** at 60 X, **b** at 200 X

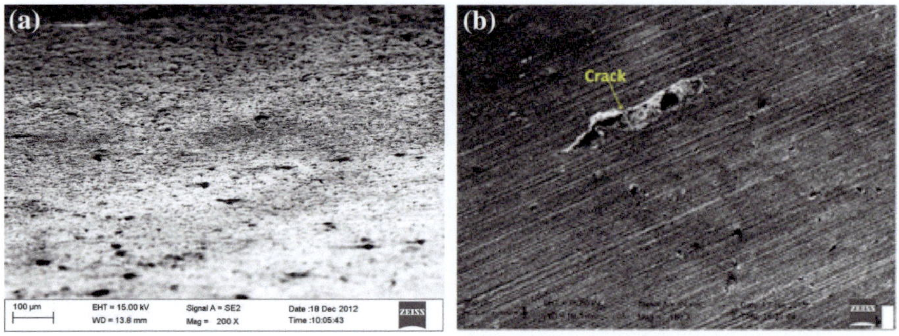

Fig. 4.12 SEM images showing **a** voids and recast layer; and **b** crack on the gear tooth surface manufactured at high discharge energy parameter settings during pilot experiments

of the recast layer, generate crack-free surface and burr-free gear tooth profile is very necessary for optimum utilization of WSEM process.

Since, SEM was unable to find the thickness of recast layer for the best quality miniature gear; therefore, micro-hardness testing was done. Results of micro-hardness testing for the best quality WSEMed miniature gear presented in the Table 4.3 and Fig. 4.13 indicate the presence of very thin (<3.5 μm) recast layer, because there is no significant difference between micro-hardness values of plain brass sample (i.e., pre-WSEM) and WSEMed gear tooth surface at different depths and no significant change in micro-hardness of the WSEMed miniature gears was observed along the depth. It seems that 170 μs pulse-off time is sufficient to expelled-out the molten material formed at 15 V and 1 μs pulse-on time. Combination of 15 V, 1 μs pulse-on time, 170 μs pulse-off time, and 9 m/min wire feed rate was found suitable for manufacturing the miniature gears having very thin recast layer.

Table 4.3 Results of micro-hardness testing for the best quality miniature gear

Outputs	Plain brass sample (Pre-WSEM)			WSEMed miniature gear tooth surface (Post-WSEM)		
Indentation force (g)	50	100	200	50	100	200
Average diagonal length (μm)	25	36	50	25	35	50
Average depth of indentation (μm)	3.5	5	7	3.5	5	7
Average micro-hardness (HV)	149	140	142	152	150	144

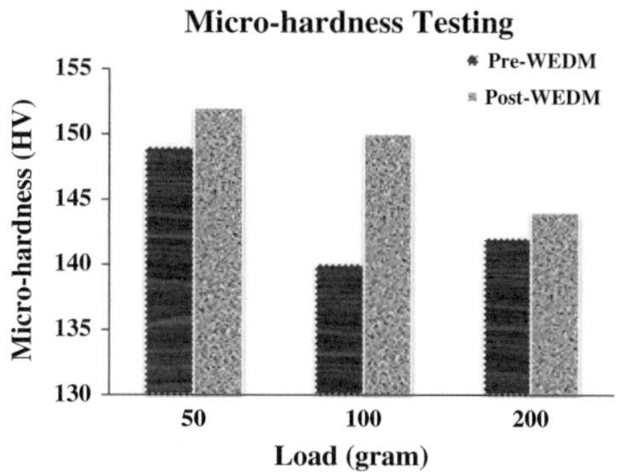

Fig. 4.13 Comparison of the micro-hardness before and after WSEM for the best quality miniature gear [9]

4.3 Conclusions from Pilot Experiments

From the analysis of the results of the pilot experiments, it was found that the main causes of micro-geometry errors are irregular shaped craters created due to violent sparks having high discharge energy and deviation of wire from its path (wire-lag). Therefore, use of low discharge energy parameters i.e., lower values of voltage and pulse-on time are recommended to achieve the high quality of miniature gears by WSEM. Moreover, longer pulse-off time leads to increase in micro-geometry errors but decrease in surface roughness; maximum value of wire feed rate is desirable for efficient and stable manufacturing of miniature gears by WSEM; and maximum cutting speed of wire spark erosion machine tool was recognized as the desirable speed to achieve high quality.

The best quality miniature gear manufactured by WSEM during pilot experiments has DIN quality number 6 for pitch (with an accumulated pitch error of 11.2 μm), and DIN quality number 8 for profile (with total profile error of 13.2 μm) besides very thin recast layer and surface defect free microstructure.

It can be concluded that WSEM is capable enough and can be explored further for manufacturing high quality miniature gears. The bracketed range of WSEM parameters for further investigations are 5–15 V for voltage, 0.6–1.0 μs for pulse-on time, 90–170 μs for pulse-off time, and 9–15 m/min for wire feed rate.

4.4 Results of Main Experiments and Their Analysis

Table 4.4 presents the 29 experimental combinations of WSEM parameters designed by Box-Behnken approach of RSM and corresponding values of F_a, F_p, R_a, R_{max} and *GCR*. Table 4.5 presents the results of ANOVA for F_a and F_p. Table 4.6 presents results of ANOVA for R_a, R_{max} and GCR. It was found by ANOVA that voltage, pulse-on time, pulse-off time and wire feed rate significantly affect all the responses. Some significant interactions which affect F_a and F_p were found between V and T_{on}, T_{on} and T_{off}. In addition to that F_a is also affected by interaction between T_{on} and W. The interaction between T_{on} and T_{off} affects the R_{max} significantly, whereas *GCR* is significantly affected by the interactions between T_{on} and W, T_{off} and W. This section describes in detail about the individual effect of four WSEM parameters on all the responses.

4.4.1 Effect of WSEM Process Parameters and Their Interactions

The effects of significant WSEM parameters on the responses were analysed and are presented graphically in terms of the experimental data points and corresponding regression equations in the following sections along with plots of the response surfaces for the significant interactions.

4.4.1.1 Effect of Pulse-on Time

Based on the experimental results given in Table 4.4, the effects of T_{on} on F_a, F_p, R_a, R_{max} and *GCR* are depicted by Fig. 4.14a–c. It was observed that increase in pulse-on time leads to increase in F_a, F_p, R_a, R_{max} and GCR. Longer pulse-on time causes excessive wire-lag due to the generation of high amount of forces by pressure of the gas bubbles and inherent to spark generation [11]. This increases the errors in total profile and accumulated pitch (as shown in Fig. 4.14a). Longer pulse-on time increases the discharge energy of the plasma channel and the time duration available for transferring this energy to the electrodes. This leads to occurrence of violent spark and improper flushing of excess amount of molten material in the machining zone which in turn form irregular shaped deeper craters and thereby deteriorates the surface finish (as shown in Fig. 4.14b) [12]. Rapid and high amount of material removal by formation of deeper and bigger craters at higher values of T_{on} resulted in high gear cutting rate (as shown in Fig. 4.14c) [13].

Table 4.4 Values of input parameters and corresponding values of the responses for different runs of main experiments

Expt. No.	Input parameters				Responses														
	V (V)	T_{on} (μs)	T_{off} (μs)	W (m/min)	'F_a' (μm)			'F_p' (μm)			'R_a' (μm)			'R_{max}' (μm)			'*GCR*' (mm^3/min)		
					R1	R2	Avg.	R1	R2	Avg.	R1	R2	Avg.	R1	R2	Avg.	R1	R2	Avg.
1	15	0.8	130	15	13.90	14.40	14.20	33.00	27.40	30.20	1.58	1.82	1.70	7.95	6.85	7.40	36.47	39.53	38.00
2	10	0.8	90	9	14.50	14.50	14.50	42.45	39.55	41.00	1.80	2.20	2.00	8.35	10.05	9.20	43.16	41.84	42.50
3	5	1.0	130	12	13.90	14.10	14.00	30.80	28.00	29.40	1.68	1.92	1.80	8.70	8.74	8.72	23.81	27.55	25.68
4	10	0.8	130	12	12.95	13.25	13.10	12.80	12.00	12.40	1.47	1.33	1.40	7.50	6.96	7.23	31.00	25.00	28.00
5	15	0.6	130	12	13.60	14.40	14.00	27.60	20.80	24.20	1.44	1.44	1.44	7.24	7.04	7.14	29.00	21.92	25.46
6	5	0.8	130	9	14.00	14.80	14.40	32.00	32.20	32.10	1.60	1.74	1.70	9.20	6.8	8.00	30.35	32.45	31.40
7	5	0.8	170	12	13.00	13.00	13.00	18.30	20.10	19.20	1.10	1.60	1.35	6.72	7.02	6.87	16.60	19.00	17.80
8	10	1.0	130	9	14.80	14.40	14.60	48.55	40.45	44.50	1.66	1.98	1.82	8.89	8.81	8.85	38.21	32.95	35.58
9	15	0.8	130	9	15.25	14.35	14.80	36.80	40.40	38.60	1.83	1.69	1.76	9.70	7.40	8.55	40.40	35.60	38.00
10	10	0.8	170	15	12.90	13.20	13.10	18.05	18.15	18.10	1.30	1.26	1.28	5.98	8.04	7.01	26.00	29.2	27.60
11	15	0.8	170	12	15.00	13.80	14.40	33.05	36.95	35.00	1.77	1.65	1.71	6.66	7.94	7.30	28.00	28.40	28.20
12	15	1.0	130	12	15.45	14.95	15.20	40.60	41.00	40.80	2.10	1.83	1.92	8.90	8.90	8.90	46.84	38.00	42.42
13	10	0.8	130	12	12.60	13.00	12.80	14.90	17.60	16.25	1.85	1.45	1.65	7.70	6.30	7.00	32.45	28.55	30.50
14	10	0.8	130	12	12.30	12.70	12.50	11.25	12.35	11.80	1.62	1.78	1.70	6.58	6.98	6.78	31.00	38.00	34.50
15	10	0.8	130	12	13.10	13.10	13.10	16.30	13.70	15.00	1.60	1.60	1.60	6.40	7.40	6.90	27.25	28.35	27.80
16	15	0.8	90	12	14.55	15.05	14.80	41.65	29.75	35.70	1.82	1.92	1.87	7.87	9.53	8.70	42.90	30.00	36.45
17	10	0.8	170	9	14.30	13.30	13.80	28.65	36.15	32.40	1.80	1.56	1.68	7.55	8.05	7.80	27.18	23.90	25.54
18	10	1.0	90	12	14.10	13.70	13.90	35.85	40.15	38.00	1.95	1.99	1.97	8.82	10.78	9.80	36.88	44.58	40.73
19	5	0.8	90	12	14.65	13.95	14.30	34.00	34.00	34.00	1.40	1.82	1.61	8.66	7.74	8.20	27.00	29.00	28.00
20	10	0.8	130	12	13.15	12.85	13.00	21.95	14.45	18.20	1.78	1.74	1.76	5.98	8.24	7.11	33.10	30.90	32.00
21	10	0.6	90	12	14.40	14.00	14.20	28.00	28.70	28.35	1.46	1.80	1.63	8.88	5.72	7.30	32.27	24.05	28.16
22	10	0.6	170	12	11.85	11.55	11.70	8.70	10.10	9.40	1.24	1.56	1.40	6.80	7.20	7.00	24.50	23.50	24.00
23	10	0.8	90	15	14.35	12.65	13.50	27.60	22.60	25.10	1.89	1.21	1.55	8.55	7.91	8.23	27.64	37.26	32.45
24	10	1.0	130	15	13.80	14.20	14.00	25.00	29.00	27.00	1.40	2.08	1.74	9.37	8.03	8.70	33.80	40.54	37.17

(continued)

Table 4.4 (continued)

Expt. No.	Input parameters				Responses														
	V (V)	T_{on} (μs)	T_{off} (μs)	W (m/min)	'F_a' (μm)			'F_p' (μm)			'R_a' (μm)			'R_{max}' (μm)			'*GCR*' (mm^3/min)		
					R1	R2	Avg.	R1	R2	Avg.	R1	R2	Avg.	R1	R2	Avg.	R1	R2	Avg.
25	5	0.6	130	12	13.00	13.60	13.30	21.25	20.05	20.65	1.17	1.10	1.14	6.82	6.62	6.72	16.16	16.00	16.08
26	10	0.6	130	9	14.05	12.95	13.50	24.00	26.00	25.00	1.50	1.51	1.49	7.22	7.68	7.45	34.18	27.10	30.64
27	10	1.0	170	12	14.20	15.00	14.60	36.85	28.75	32.80	1.86	1.44	1.65	7.90	7.90	7.90	24.00	33.6	28.80
28	5	0.8	130	15	12.90	13.10	13.00	26.70	18.10	22.40	1.38	1.12	1.25	6.32	7.18	6.75	21.48	22.72	22.10
29	10	0.6	130	15	12.25	11.75	12.00	17.20	14.80	16.00	1.18	1.18	1.18	5.80	8.00	6.90	19.85	20.67	20.26

Table 4.5 Results of ANOVA for F_a and F_p [11]

Source	Sum of squares	DF	Mean square	F value	P value Prob > F	Remarks
Total profile error 'F_a'						
Model	20.43	14	1.46	34.43	<0.0001	Highly significant
V	2.43	1	2.43	57.34	<0.0001	Highly significant
T_{on}	4.81	1	4.81	113.57	<0.0001	Highly significant
T_{off}	1.76	1	1.76	41.61	<0.0001	Highly significant
W	2.80	1	2.80	66.15	<0.0001	Highly significant
V^2	4.78	1	4.78	112.76	<0.0001	Highly significant
T_{on}^2	0.62	1	0.62	14.55	0.0019	Significant
T_{off}^2	1.08	1	1.08	25.52	0.0002	Significant
W^2	0.83	1	0.83	19.65	0.0006	Significant
$V \times T_{off}$	0.20	1	0.20	4.78	0.0463	Significant
$T_{on} \times T_{off}$	2.56	1	2.56	60.40	<0.0001	Highly significant
$T_{on} \times W$	0.20	1	0.20	4.78	0.0463	Significant
Residual	0.59	14	0.042			
Lack of fit	0.33	10	0.033	0.51	0.8214	Insignificant
Pure error	0.26	4	0.065			
Total	21.02	28				
R-Squared = 0.972 Adjusted R-Squared = 0.943 Predicted R-Squared = 0.889 Adequate Precision = 23.471 Std. Deviation = 0.21						
Accumulated pitch error 'F_p'						
Model	2631.88	14	187.99	34.25	<0.0001	Highly significant
V	182.13	1	182.13	33.34	<0.0001	Highly significant
T_{on}	675.00	1	675.00	123.55	<0.0001	Highly significant
T_{off}	265.36	1	265.36	48.57	<0.0001	Highly significant
W	465.26	1	465.26	85.16	<0.0001	Highly significant

(continued)

Table 4.5 (continued)

Source	Sum of squares	DF	Mean square	F value	P value Prob > F	Remarks
V^2	501.89	1	501.89	91.86	<0.0001	Highly significant
T_{on}^2	188.45	1	188.45	34.49	<0.0001	Highly significant
T_{off}^2	319.54	1	319.54	58.49	<0.0001	Highly significant
W^2	372.20	1	372.20	68.12	<0.0001	Highly significant
$V \times T_{off}$	49.70	1	49.70	9.10	0.0093	Significant
$T_{on} \times T_{off}$	55.13	1	55.13	10.09	0.0067	Significant
Residual	76.49	14	5.46			
Lack of fit	48.05	10	4.81	0.68	0.7203	Insignificant
Pure error	28.44	4	7.11			
Total	2708.61	28				

R-Squared = 0.971 Adjusted R-Squared = 0.943 Predicted R-Squared = 0.880
Adequate Precision = 19.182 Std. Deviation = 2.34

Table 4.6 Results of ANOVA for R_a, R_{max} [12] and *GCR* [13]

Source	Sum of Squares	DF	Mean Square	F Value	P value Prob > F	Remarks
Average surface roughness 'R_a'						
Model	1.23	4	0.31	29.88	<0.0001	Highly significant
V	0.20	1	0.20	19.45	0.0002	Significant
T_{on}	0.57	1	0.57	55.58	<0.0001	Highly significant
T_{off}	0.20	1	0.20	19.70	0.0002	Significant
W	0.26	1	0.26	24.79	<0.0001	Highly significant
Residual	0.25	24	0.01			
Lack of fit	0.17	20	0.008	0.45	0.895	Insignificant
Pure error	0.076	4	0.019			
Total	1.48	28				

R-Squared = 0.832 Adjusted R-Squared = 0.804 Predicted R-Squared = 0.763
Adequate Precision = 17.289 Std. Deviation = 0.1

(continued)

Table 4.6 (continued)

Source	Sum of Squares	DF	Mean Square	F Value	P value Prob > F	Remarks
Maximum surface roughness 'R_{max}'						
Model	20.61	14	1.472	30.041	<0.0001	Highly significant
V	0.621	1	0.621	12.669	0.0031	Significant
T_{on}	8.944	1	8.944	182.450	<0.0001	Highly significant
T_{off}	4.750	1	4.750	96.903	<0.0001	Highly significant
W	1.968	1	1.968	40.153	<0.0001	Highly significant
V^2	0.448	1	0.448	9.152	0.0091	Significant
$T_{on}{}^2$	1.816	1	1.816	37.06	<0.0001	Highly significant
$T_{off}{}^2$	1.757	1	1.757	35.84	<0.0001	Highly significant
W^2	1.383	1	1.383	28.21	0.0001	Significant
T_{on} x T_{off}	0.640	1	0.640	13.05	0.002	Significant
Residual	0.686	14	0.049			
Lack of fit	0.562	10	0.056	1.826	0.294	Insignificant
Pure error	0.123	4	0.030			
Total	21.303	28				
R-Squared = 0.967 Adjusted R-Squared = 0.935 Predicted R-Squared = 0.838 Adequate Precision = 20.847 Std. Deviation = 0.22						
Gear cutting rate '*GCR*'						
Model	1258.37	14	89.88	18.03	<0.0001	Highly significant
V	379.91	1	379.91	76.22	<0.0001	Highly significant
T_{on}	360.58	1	360.58	72.34	<0.0001	Highly significant
T_{off}	264.14	1	264.14	52.99	<0.0001	Highly significant
W	57.12	1	57.12	11.46	0.0044	Significant
W^2	34.10	1	34.10	6.84	0.0203	Significant
T_{on} x W	35.82	1	35.82	7.19	0.0179	Significant
T_{off} x W	36.66	1	36.66	7.36	0.0169	Significant
Residual	69.78	14	4.98			
Lack of fit	41.75	10	4.18	0.60	0.7694	Insignificant
Pure error	28.03	4	7.01			
Total	1328.15	28				
R-Squared = 0.947 Adjusted R-Squared = 0.894 Predicted R-Squared = 0.786 Adequate Precision = 15.372 Std. Deviation = 2.23						

4.4.1.2 Effect of Pulse-Off Time

The effects of pulse-off time on all response parameters are depicted by Fig. 4.15. It is evident from Fig. 4.15a that with increase in the pulse-off time, both F_a and F_p initially decrease; attain minimum values and then start increasing again i.e., optimum range of pulse-off time (150–170 μs) exists for minimum values of F_a and F_p.

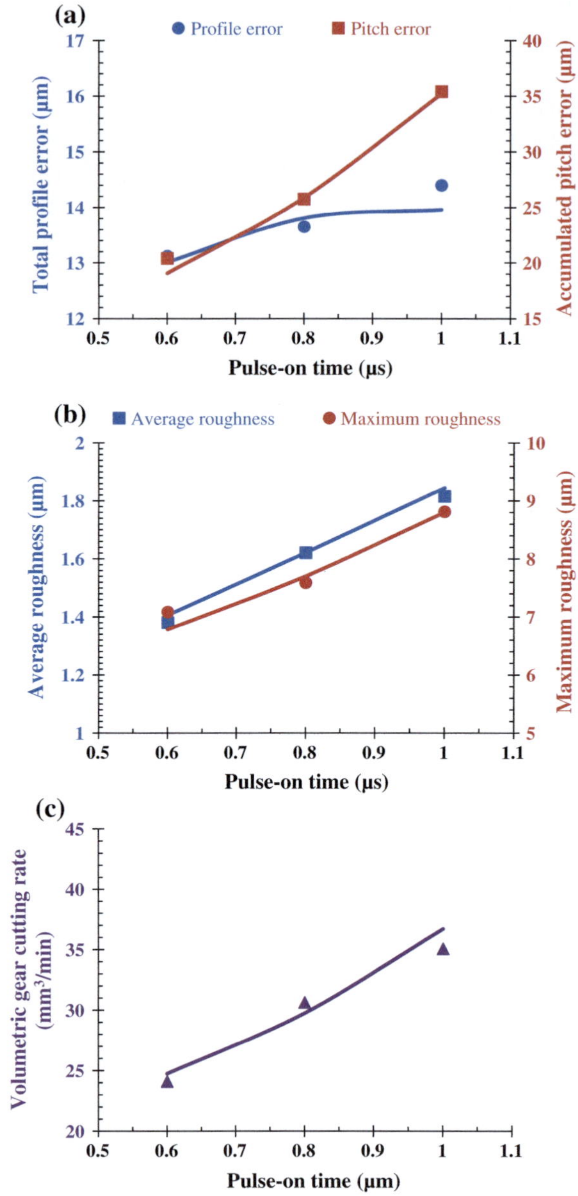

Fig. 4.14 Effect of pulse-on time on **a** F_a and F_p [11]; **b** R_a and R_{max} [12]; and **c** GCR [13]

Because, a shorter pulse-off time causes incomplete removal of the machined material from the gap between the wire and the gear. This residual debris causes short-circuit and vibrations in the wire. These vibrations cause wire lag which finally produces more errors in profile and pitch of the gear [11]. Slightly longer pulse-off time minimizes the wire vibrations by reducing the chances of the

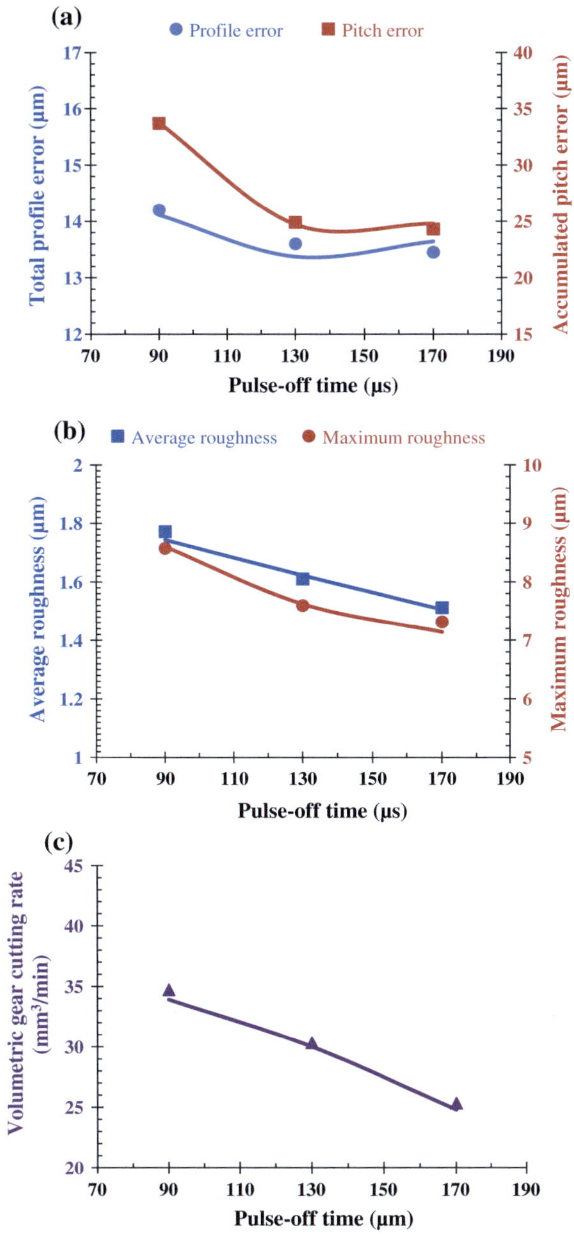

Fig. 4.15 Effect of pulse-off time on **a** F_a and F_p [11]; **b** R_a and R_{max} [12]; and **c** GCR [13]

short-circuiting, but after a certain limit longer pulse-off time increases the amount of hydraulic forces generated by dielectric flushing [2, 11]. These forces again increase F_a and F_p by producing wire-lag.

Figure 4.15b shows that with increase in pulse-off time R_a and R_{max} decreases almost linearly. While increase in surface roughness with lower pulse-off time is due to (i) poor flushing of the molten material from the machining zone, (ii) short-circuiting of wire electrode due to the presence of residual or un-expelled molten material in the inter-electrode gap [12]. Gear cutting rate is highest at the lowest value of pulse-off time and decreases with increase in it (Fig. 4.15c).

4.4.1.3 Effect of Voltage

The variation of response parameters with voltage is depicted by Fig. 4.16a–c. Both F_a and F_p initially decrease with increase in voltage, attain the minimum values and then start increasing again (as shown in Fig. 4.16a) i.e., optimum range of voltage (8–9 V) exists to minimize F_a and F_p. This can be explained by the fact that in WSEM, a smaller voltage requires a closer gap between wire and workpiece to ensure the discharge of spark. But, this proximity increases the electrostatic force (the force due to movement of ions) between the wire and workpiece causing the wire deflection which produces more deviations in profile and pitch [11]. Machining at higher voltage does not need a closer gap due to stronger electric field. At higher voltage the discharge energy is high which leads to the formation of irregular shaped deeper craters, and the generation of the forces due to violent spark and the pressure of the gas bubbles. These forces produced comparatively higher deviations in the position of the wire and thereby more deviations or errors in profile and pitch. As the amount of wire lag is minimum in the range of 8–9 V so, the minimum values of F_a and F_p were obtained in this range. Increase in voltage leads to increase the discharge energy which in turn generates violent spark and thereby irregular shaped deeper craters are formed [12]. This consequently increases the values of R_a and R_{max} (as shown in Fig. 4.16b). On the other hand stronger electric field at higher voltage facilitates the ionization of dielectric and thereby increases the occurrence of discharge, which results in melting and evaporation of high amount of material and consequently enhances the gear cutting rate (as shown in Fig. 4.16c) [13].

4.4.1.4 Effect of Wire Feed Rate

The significant effects of wire feed rate on the response parameters are shown by Fig. 4.17a–c. The values of F_a and F_p are highest at lowest value of the wire feed rate and as the feed rate increases the F_a decreases and F_p starts decreasing up-to a certain value and then start increasing again (refer Fig. 4.17a). The optimum range (12–15 m/min) of wire feed rate exists for minimum values of F_a and F_p. This can be explained by the fact that low wire feed rate exposes the wire to bear more discharge energy per unit wire length which increases wire-lag, wire-breakage and non-uniform wear of wire. All these factors lead to higher F_a and F_p [11]. Increase in wire feed rate reduces the duration of spark concentration at a particular location

Fig. 4.16 Effect of voltage on **a** F_a and F_p [11]; **b** R_a and R_{max} [12]; and **c** GCR [13]

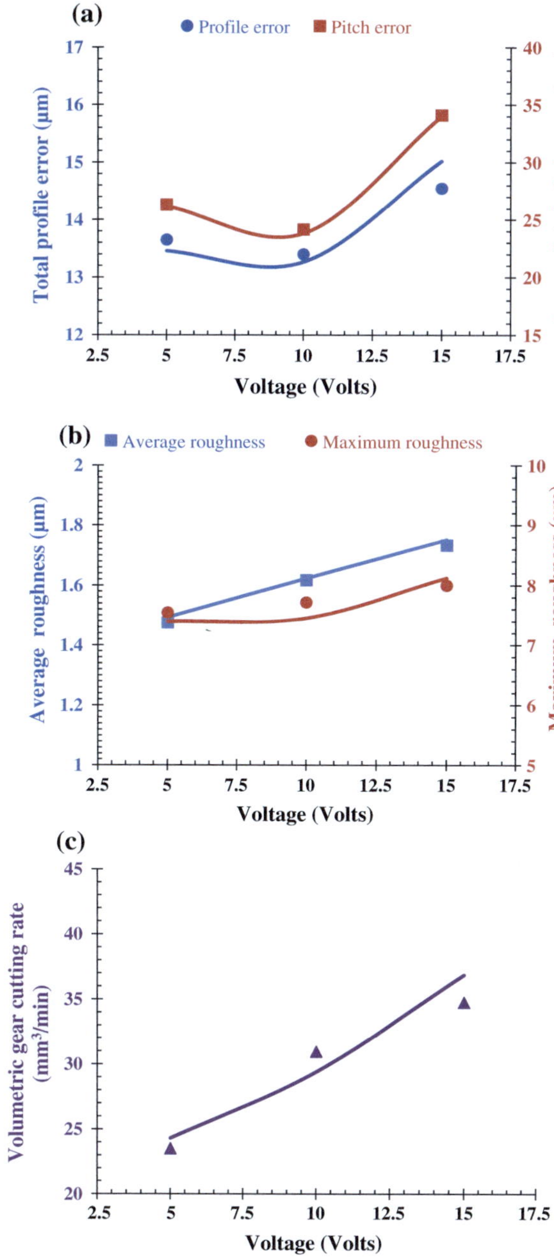

on the wire, thereby decreasing wire-lag and wire-breakage and consequently results in the lower values of F_a and F_p. But, a very high wire feed rate makes the fresh wire available at a faster rate which increases conductance of wire causing more spark energy. This again increases F_p.

It is evident from the Fig. 4.17b that with increase in wire feed rate R_a decreases almost linearly, while R_{max} decreases non-linearly with occurrence of optimum

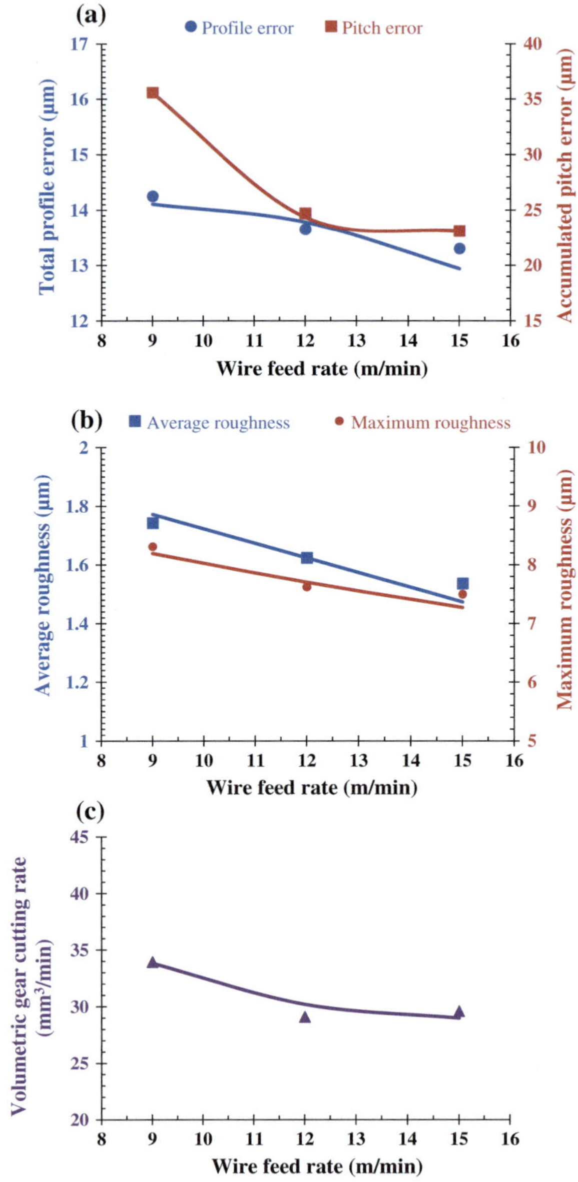

Fig. 4.17 Effect of wire feed rate on **a** F_a and F_p [11]; **b** R_a and R_{max} [12]; and **c** GCR [13]

range of wire feed rate (12–15 m/min) to minimize the R_{max}. Increase in wire feed rate decreases the period of availability of wire for machining and thereby reduces the chances of adherence of wire on the gear teeth surfaces which in turn minimizes the surface roughness [12].

As illustrated in Fig. 4.17c that at lower wire feed rate *GCR* is high because the period of availability of wire is more. But, it decreases with increase in feed rate up-to a certain level then starts increasing again, as very high wire feed enhances the participation of the fresh wire for machining and thereby increases the erosion rate which in turn increases the GCR [13].

4.4.1.5 Effects of Significant Interactions

Some significant interactions between WSEM parameters affecting the responses were found by the ANOVA study. These interaction effects were plotted using the values of the responses predicted by regression analysis (Figs. 4.18, 4.19, 4.20, and 4.21). Figure 4.18a, b depicts the effect of interaction between voltage and pulse-off time on F_a and F_p, whereas Fig. 4.19a, b depicts the variation in F_a and F_p with interaction between pulse-on time and pulse-off time. It is evident from both the figures that the increase either in voltage or pulse-on time, which is required to increase the material removal rate, necessitates increase in pulse-off time in order to keep lower values of F_a and F_p but, up to only a certain level, because very high pulse-off time may again cause higher values of F_a and F_p. However, Fig. 4.19a suggests the combination of lower pulse-on and higher pulse-off for minimizing the F_a [11].

The only significant interaction found to affect the maximum roughness was the interaction between the pulse-on time and pulse-off time (Fig. 4.19c) indicates that for an increase in the pulse-on time (which is required to get more material removal rate), the pulse-off time should also be increased so that the maximum surface roughness remains at the same value [12].

Effect of interaction between pulse-on time and wire feed rate on F_a and *GCR* is shown by Fig. 4.20a, b. Figure 4.20a suggests the use of shorter pulse-on time and

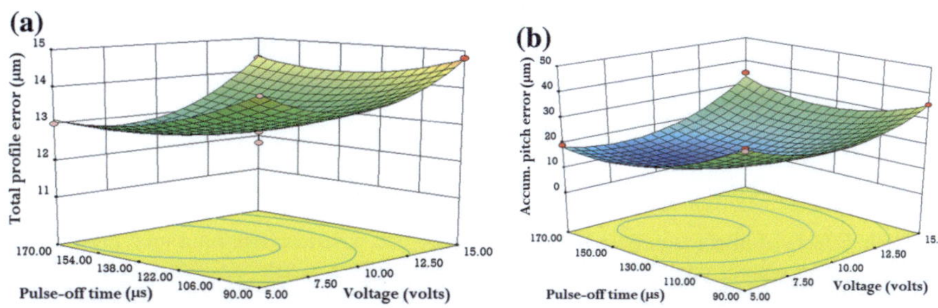

Fig. 4.18 Effects of interaction between voltage and pulse-off time on **a** total profile error 'Fa' and **b** accumulated pitch error 'F_p' of WSEMed miniature gears [11]

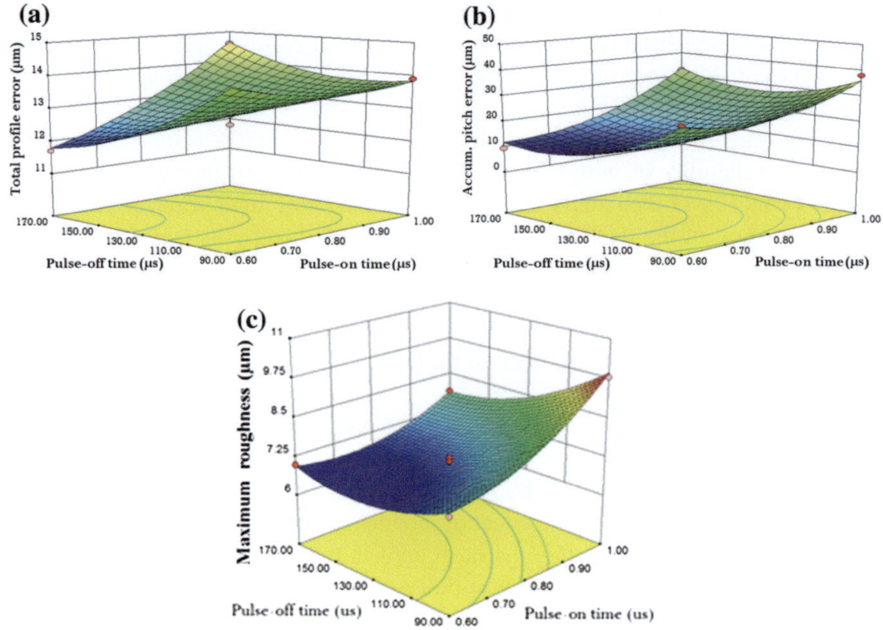

Fig. 4.19 Effects of interaction between pulse-on time and pulse-off time on **a** total profile error 'F_a'; **b** accumulated pitch error 'F_p'; **c** Maximum roughness 'R_{max}' of WSEMed miniature gears

higher wire feed rate for smaller F_a. On the other hand the increase in pulse-on time for increasing *GCR* needs simultaneous increment in wire feed rate (Fig. 4.20b). Interaction between pulse-off time and wire feed rate (Fig. 4.21) suggests that increase in pulse-off time for better surface characteristics needs simultaneous decrease in wire feed rate in order to maintain the *GCR* [13].

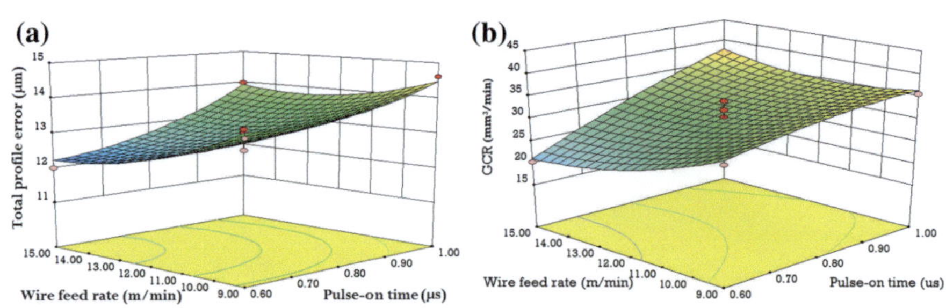

Fig. 4.20 Effects of interaction between pulse-on time and wire feed rate on **a** total profile error 'F_a' [11]; **b** gear cutting rate '*GCR*' [13] of WSEMed miniature gears

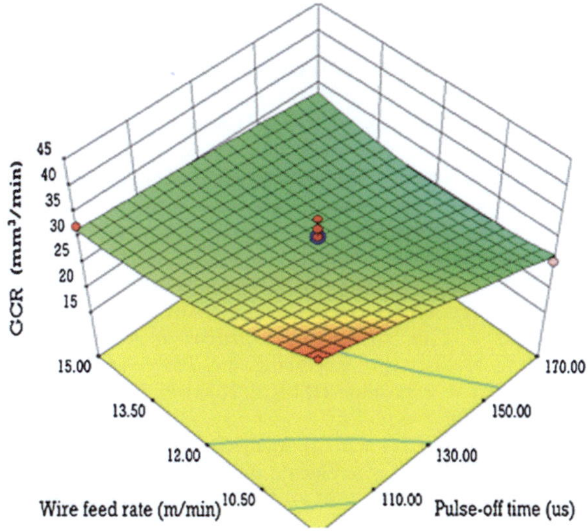

Fig. 4.21 Effect of interaction between pulse-off time and wire feed rate on gear cutting rate '*GCR*' [13]

4.5 Conclusions from Main Experiments

From the analysis of the results of main experiments, voltage; pulse-on time; pulse-off time and wire feed rate were found to be highly significant parameters affecting the micro-geometry, surface roughness and gear cutting rate. It was also observed that longer pulse-on time increases the gear cutting rate but deteriorates micro-geometry and surface finish, increase in voltage deteriorates the surface finish but increases the gear cutting rate, optimum range (150–170 μs) of pulse-off time exists to minimize F_a and F_p, increase in pulse-off time leads to decrease in the gear cutting rate and improvement in the surface finish, optimum range of voltage (8–9 V) exists to minimize F_a and F_p, optimum range of wire feed rate (12–15 m/min) exists to minimize F_a; F_p and R_{max} and to maximize *GCR*.

Consequently, it was concluded that the trends are conflicting in nature, a trade-off exists and optimization of WSEM is necessary for simultaneous minimization of micro-geometry errors and surface roughness, and maximization of gear cutting rate.

References

1. Gupta K, Chaubey SK, Jain NK (2014) Exploring wire-EDM for manufacturing the high quality meso-gears. Proc Mater Sci 5:1755–1760
2. Okada A, Konishi T, Okamoto Y, Kurihara H (2015) Wire breakage and deflection caused by nozzle jet flushing in wire EDM. Ann CIRP 64:233–236
3. Gupta K, Jain NK (2013) On micro-geometry of miniature gears manufactured by WEDM. Mater Manuf Processes 28(10):1153–1159

4. Arunachalam C, Aulia M, Bozkurt B, Eubank PT (2001) Wire vibration, bowing, and breakage in wire electrical discharge machining. J Appl Phys 89(8):4255–4262
5. Mingqi L, Minghui L, Guangyao X (2005) Study on the variations of form and position of the wire electrode in WSEM-HS. Int J Adv Manuf Technol 25:929–934
6. Puri AB, Bhattacharyya B (2003) An analysis and optimization of the geometrical inaccuracy due to wire lag phenomenon in WSEM. Int J Mach Tool Manuf 43:151–159
7. Sarkar S, Sekh M, Bhattacharyya B (2011) A novel method of determination of wire lag for enhanced profile accuracy in WSEM. Precis Eng 35:339–347
8. Liao YS, Huang JT, Chen YH (2004) A study to achieve fine surface finish in Wire-SEM. J Mater Process Technol 149:165–171
9. Gupta K, Jain NK (2014) On surface integrity of miniature spur gears manufactured by wire electrical discharge machining. Int J Adv Manuf Technol 72(9–12):1735–1745
10. Gupta K, Jain NK (2013) Deviations in geometry of miniature gears fabricated by wire electrical discharge machining. In: Proceedings of international mechanical engineering congress & exposition (IMECE 2013) of ASME, V010T11A047, San Diego, USA, 13–21 Nov. 2013
11. Gupta K, Jain NK (2014) Analysis and optimization of micro-geometry of miniature gears manufactured by wire electric discharge machining. Precis Eng 38(4):728–737
12. Gupta K, Jain NK (2014) Analysis and optimization of surface finish of wire electrical discharge machined miniature gears. Proc IMechE B J Eng Manuf 228(5):673–681
13. Gupta K, Jain NK (2013) On productivity of wire electric discharge machining for manufacturing of miniature gears. In: Proceedings of 2nd international conference on intelligent robotics, automation and manufacturing 2013 (IRAM 2013), Indore, India, 16–18 Dec. 2013, pp 428–439

Chapter 5
Modelling and Optimization

5.1 Modelling of WSEM Process

Understanding the dependency or relation of performance measures (output parameters/responses) with input parameters is essential in order to improve the process performance of WSEM. WSEM process needs to be modelled for the same. The success of optimization techniques also depends on the establishment of proper relationships between input parameters and performance measures. But, due to the stochastic and complex nature of WSEM process, it is too difficult to establish such relationships. During earlier days, physical models of WSEM process were developed based on the actual mechanism. Since, the physical models could not establish the relationships between input parameters and the performance measures accurately, as the process involves many thermal, electrical and metallurgical variables. Further, inevitable assumptions made while developing physical models of the process, induce large deviations from actual process. Thus, inability of physical models has led the researchers to develop empirical or data based models of WSEM process. In the present work, the relations between WSEM process parameters and responses have been developed by two approaches namely regression analysis and artificial neural network (ANN). A detailed description of these approaches is given in the following sections.

5.1.1 Modelling Using Regression Analysis

Regression analysis is a method to determine the relationship between a set of independent variables and a response in terms of a mathematical model. It is a general approach to fit empirical models for the process to be studied and applied to gain insight and understanding about it. These mathematical models could be used for interpolation i.e. for predicting the response at the factor levels between those

© Springer Science+Business Media Singapore 2017

K. Gupta and N.K. Jain, *Near-Net Shape Manufacturing of Miniature Spur Gears by Wire Spark Erosion Machining*, Materials Forming, Machining and Tribology, DOI 10.1007/978-981-10-1563-2_5

actually used in the experiment. The relationship between the response variable and a set of independent variables is usually unknown. In general, the low-order polynomial model is used to develop this relationship using least square method of curve fitting. A polynomial model sufficiently approximates a small region of the response surface. Therefore, depending on the approximation of unknown function *f*, either first-order or second-order models are employed.

First Order Model

A first-order model uses low-order polynomial terms to describe some part of the response surface. This model is appropriate for describing a flat surface with or without tilted surfaces. A first-order model of a response $\boldsymbol{y}$ function of $\boldsymbol{q}$ independent variables $\boldsymbol{x_1}, \boldsymbol{x_2}, \ldots, \boldsymbol{x_q}$ can be expressed as follows [1]:

$$y = \beta_0 + \beta_1 x_1 + \beta_2 x_2 + \ldots + \beta_q x_q + \varepsilon \tag{5.1}$$

Where, ε is the experimental error;

The parameters β_j $(j = 0, 1, 2, \ldots, q)$ are the regression coefficients and measure the expected change in response y per unit increase in x_j when the other independent variables are held constant. When there are more than two independent variables, the first order regression model is called *multiple-regression model.*

Second Order Model

When the response has non-linear dependence on the independent variables i.e. the response surface has curvature then the first-order model is insufficient to describe it and a second or higher order model is useful in approximating the true response surface with parabolic curvature. The second-order model includes all the terms in the first-order model, plus all quadratic terms and all interaction terms. It is usually expressed as:

$$y = \beta_0 + \sum_{j=1}^{q} \beta_j x_j + \sum_{j=1}^{q} \beta_{jj} x_j^2 + \sum_{j=}^{q} \sum_{i<j} \beta_{ij} x_i x_j + \varepsilon \tag{5.2}$$

The second order model is flexible, because it can take a variety of functional forms and approximates the response surface locally. Therefore, this model is usually a good estimation of the true response surface.

In the present work regression modeling for each individual response was done by using the software *Design Expert 8*. For average roughness, linear model was found fit to model its response surface whereas, for total profile error, accumulated pitch error, maximum roughness and GCR quadratic (second order) models were found to suitable to model their responses. Based on the ANOVA study, these models were then restructured to include only the significant parameter and their interaction terms. All these models with details of adequacy testing and residual analysis are described in the following sections. Various important terms of ANOVA and model adequacy testing have already been discussed in the Chap. 3.

5.1.1.1 Regression Models for Micro-geometry

Quadratic models were found suitable for 'F_a' and 'F_p'. Equations 5.3 and 5.4 represent regression models for 'F_a' and 'F_p' in terms of significant parameters and interactions.

$$F_a = 50.1 - 1.003V - 27.92T_{on} - 0.17T_{off} - 1.63W + 0.03V^2 + 7.71T_{on}^2 + 0.00025T_{off}^2 + 0.04W^2 + 0.001VT_{off} + 0.1T_{on}T_{off} + 0.38T_{on}W \tag{5.3}$$

$$F_p = 394.45 - 10.38V - 215.69T_{on} - 1.84T_{off} - 20.08W + 0.35V^2 + 134.83T_{on}^2 + 0.004T_{off}^2 + 0.84W^2 + 0.02VT_{off} + 0.46T_{on}T_{off} \tag{5.4}$$

The residual analysis was used as the primary diagnostic tool. Normal probability plots of residuals have been drawn for both F_a and F_p as shown in Fig. 5.1. These plots clearly indicate that errors are normally distributed as most of the

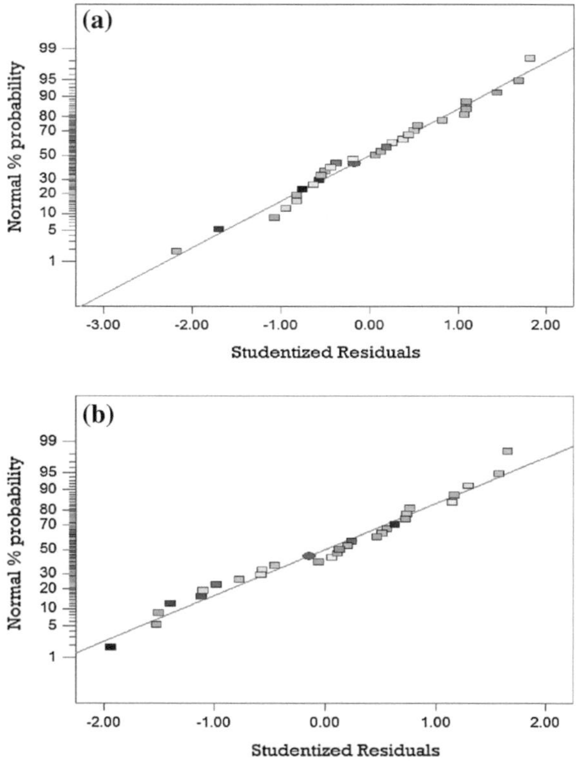

Fig. 5.1 Normal plot of residuals for **a** total profile deviation 'F_a'; and **b** accumulated pitch deviation 'F_p' [10]

residuals are clustered around a straight line. The insignificant values of *lack of fit* i.e. p values of the model for F_a (0.8214) and F_p (0.7203), confirmed the adequacy of the quadratic model. Another test, *model summary statistics* further confirmed that the quadratic model is the best fit as it exhibits low standard deviation 0.21 for F_a; 2.34 for F_p, PRESS values 2.33 for F_a and 323.29 for F_p, high "R-Squared" values (0.9718 for F_a and 0.9716 for F_p) which are closed to 1. The values of predicted R-Squared and adjusted R-Squared for both F_a and F_p are in good agreement, which indicates a high correlation between observed and predicted values. The values of adequate precision 23.47 and 19.18 respectively for F_a and F_p suggest that the quadratic models are a good predictor for them.

5.1.1.2 Regression Models for Surface Roughness

The trends of variation of 'R_a' with WSEM parameters were linear as mentioned in the Chap. 4. Consequently, linear model was found to be best suited to fit the experimental data for it. Normal probability plots of residuals for both R_a and R_{max} (Fig. 5.2) suggest the normal distribution of the data. Equations 5.5 and 5.6 represent these models for R_a and R_{max} respectively.

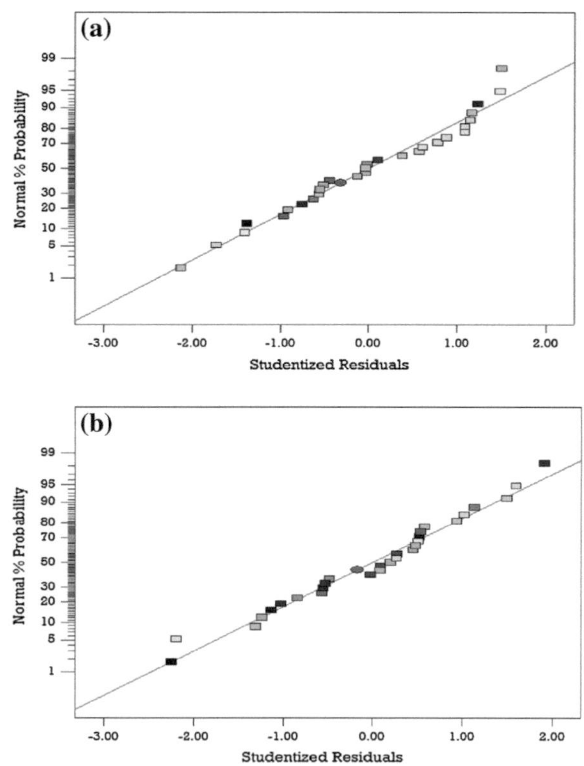

Fig. 5.2 Normal plot of residuals for **a** average roughness 'R_a'; and **b** maximum roughness 'R_{max}'

$$R_a = 1.486 + 0.026V + 1.09T_{on} - 0.003T_{off} - 0.05W \quad (5.5)$$

$$\begin{aligned} R_{max} = {} & 25.758 - 0.126V - 11.753T_{on} - 0.064T_{off} - 1.565W + 0.01V^2 \\ & + 13.231T_{on}^2 + 0.0003253T_{off}^2 + 0.051W^2 - 0.05T_{on}T_{off} \end{aligned} \quad (5.6)$$

The values of the parameters lack of fit (p value): 0.895 and 0.294; PRESS: 0.35 and 3.44; R-Squared: 0.8328 and 0.9678; predicted R-Squared: 0.7635 and 0.8387; adjusted R-Squared: 0.8049 and 0.9356; adequate precision: 17.28 and 20.847 confirm the adequacy and fitness of the models for R_a and R_{max}.

5.1.1.3 Regression Model for Gear Cutting Rate

Quadratic model (described by Eq. 5.7) was found to be sufficiently modeled the gear cutting rate (GCR). Linearity of the normal plots confirms the normal distribution of the data (Fig. 5.3) for this model.

$$\begin{aligned} GCR = {} & 30.46 + 5.63 \times V + 5.48 \times T_{on} - 4.69 \times T_{off} - 2.18W \\ & + 2.99 \times T_{on} \times W + 3.03 \times T_{off} \times W + 2.29 \times W^2 \end{aligned} \quad (5.7)$$

The values of the parameters of adequacy testing and fitness for this model are: lack of fit (p value): 0.769; PRESS: 284.28; R-Squared: 0.9475; predicted R-Squared: 0.786; adjusted R-Squared: 0.8949; adequate precision: 15.372.

5.1.2 Modeling Using Artificial Neural Network (ANN)

Sometimes regression techniques do not provide satisfactory results because of the presence of noise in the system variables of the processes. Artificial Neural

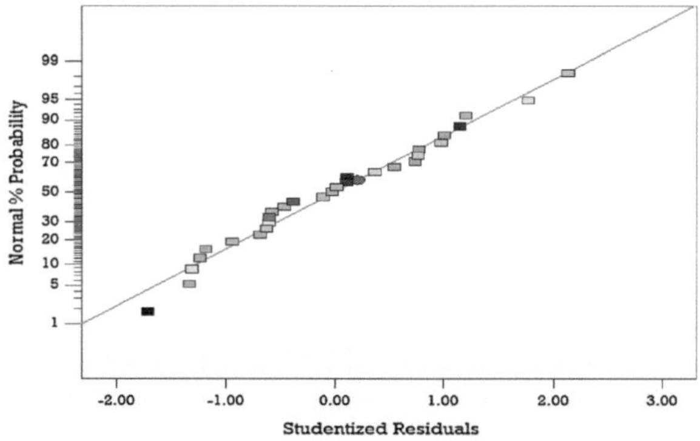

Fig. 5.3 Normal plot of residuals for *GCR*

Network (ANN) forms an alternative approach to model the system efficiently as it is capable of mapping input variable and performance measures of WSEM and hence overcomes the disadvantages of the statistical techniques. In past, several researchers have shown the applicability and superiority of ANN in modelling of WSEM [2–4]. Hence, it was also attempted to develop the ANN models in the present work.

Artificial neural networks (ANN) are simplified models of biological nervous system inspired by the computing capability of human brain [5]. ANNs have the capability to learn and thereby acquire knowledge, and make it available for use. These models are built by connecting processing units, called nodes or neurons. Each of the input is associated with some weight which takes a portion of the input to the node for processing. The node combines the inputs and produces net input which in turn trained by suitable algorithm and transformed into output with the help of transfer or activation function. ANNs are broadly classified into feed forward and recurrent networks. Feed forward networks are those in which computation flows from the input nodes to the output nodes in a sequence. The layers in between the input and output layer are known as hidden layers. The feed forward networks typically do regression task i.e. function approximation. In a recurrent network, signals may propagate from the output of any neuron to the input of any neuron.

Primarily there are two types of learning methods: supervised and unsupervised learning. In the supervised learning method, every input pattern that is used to train the network is associated with an output pattern (i.e. the target or desired pattern). A teacher is assumed to be present during the learning process to determine the error after comparing the networks computed output and desired output. Many researchers have proposed that multilayered networks are capable of computing a wide range of Boolean functions than networks with a single layer of computing units [6]. However, the computational effort needed for modeling a system increases substantially when more parameters and more complicated topologies are considered. The back propagation algorithm is found most suitable for handling such large learning problems. It is one of the most studied and used algorithm for neural networks learning ever. In feed forward networks nodes are computing units and numerical information are transmitted from node to node through connections. The network represents a series of function compositions which transform an input to an output. The learning problem consists of finding the optimal combination of weights so that network function approximates a given function as closely as possible.

BPNN applies the error-back propagation procedure for learning. The back-propagation procedure uses a gradient descent method, which adjusts the weight in its original and simplest form by an amount proportional to the partial derivative of the error function with respect to the given weight. The error-back propagation is typical supervised learning procedure. They are the most popular neural network models in EDM based processes.

Selection of network architecture has always been a big challenge in the field of neural network modelling. BP can be applied to any number of hidden layers but

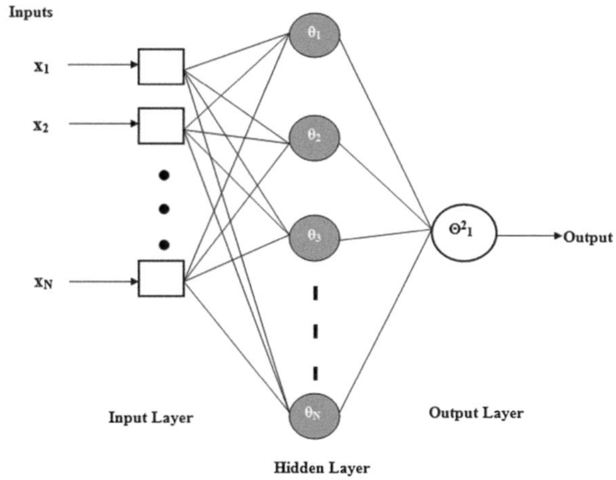

Fig. 5.4 Schematic representation of three layer BPNN

generally a three layers network is used. Three layered network means input-hidden-output structure (Fig. 5.4). Network architecture is decided by the number of inputs, number of outputs, the level of complexity of problem, the training algorithms and the level of accuracy required. Too small a number of hidden neuron may not train the network properly (under fitting) and too many hidden neurons may require excessive computer time and prevent the network from generalizing (over fitting).

In the present work, feed forward BPNN was used to develop the ANN models to develop the relationship between WSEM parameters and responses, because it has the advantages of fast response and high learning accuracy. For designing the networks, Neural Network Toolbox of MATLAB 7.7 was used. The log-sigmoid transfer function was used as the activation function for the hidden layers and for the output layers. Simultaneously, the *Levenberg–Marquardt* reduction scheme was selected for the neural network algorithm. The total number of data set is 29. The whole data set was divided into 3 sets viz. training, validation and test data set in the proportion of 60:20:20. The training data set was used to fit the model or to establish the input-output mapping. The validation data set was used to stop the training by early stopping criteria. Test data set was used to evaluate the performance. The training of the neural network model was carried out using 60 % (17 sets) of experimental data. A number of networks were tried by changing the number of neurons in the hidden layer for fastest convergence of the error and for achieving a lowest value of mean square error (*MSE*) given by Eq. (5.8):

$$MSE = \frac{1}{n}\sum_{i=1}^{n}(E_i - N_i)^2 \tag{5.8}$$

Where, n is number of experiments; E_i is experimental values of the responses; and N_i is the neural network predicted values of the responses. When the training met the performance goal (i.e. attains the minimum value of mean square error), it was then tested by using last 20 % (six different sets) of the experimental data which were not used in training the model. One important parameter *co-relation coefficient* was also used to determine how well the network output fits the desired output. Closer its value to 1 better the predictions are.

5.1.2.1 ANN Model for Micro-geometry

The multilayer feed-forward neural network developed for prediction of F_a and F_p has one input layer having four neurons corresponding to four WSEM input parameters, one hidden layer, and one output layer having two neurons corresponding to two output parameters. After running various rounds of trial and error, the best results were obtained with 14 neurons in hidden layer. Figure 5.5 presents the architecture of the developed ANN model for F_a and F_p. The value of co-relation coefficient for the developed neural network is 0.991.

5.1.2.2 ANN Model for Surface Roughness

The multilayer feed-forward neural network developed for prediction of R_a and R_{max} simultaneously has one input layer having four neurons corresponding to four WSEM input parameters, one hidden layer, and one output layer having two neurons corresponding to two output parameters. After running various rounds of trial and error, the best results were obtained with 10 neurons in hidden layer.

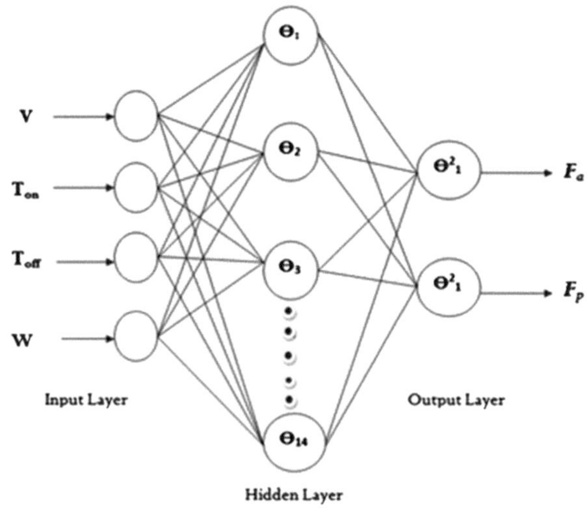

Fig. 5.5 Architecture of the developed neural network model for F_a and F_p

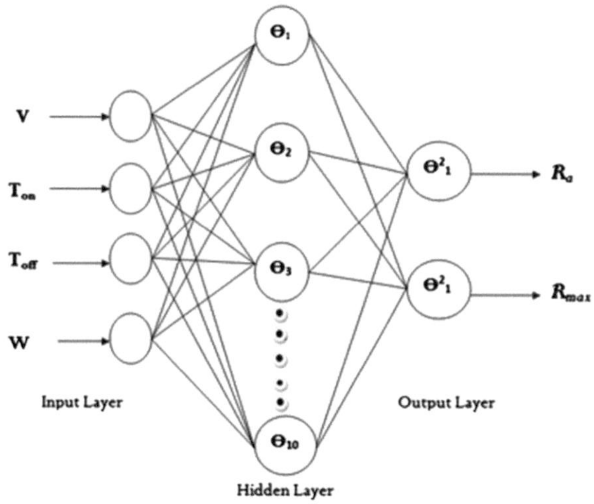

Fig. 5.6 Architecture of the developed neural network model for R_a and R_{max} [11]

Figure 5.6 shows the architecture of the developed ANN model. The value of co-relation coefficient is 0.998.

5.1.2.3 ANN Model for Gear Cutting Rate

The multilayer feed-forward neural network has one input layer having four neurons corresponding to four input parameters, one hidden layer, and one output layer having one neuron corresponding to one output parameters i.e. *GCR*. After running various rounds of trial and error, the closest results were obtained with 12 neurons

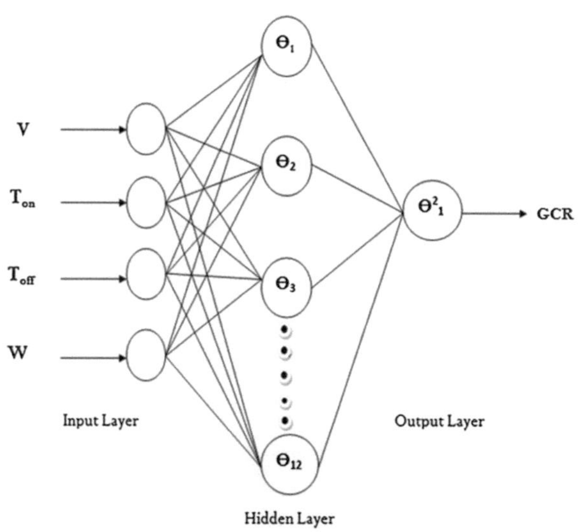

Fig. 5.7 Architecture of the developed neural network model for *GCR*

Table 5.1 Details of parameters for the developed ANN models

Responses	No. of neurons in input layer	No. of neurons in output layer	No. of neurons in hidden layer	Co-relation coefficient
F_a–F_p	4	2	14	0.991
R_a–R_{max}	4	2	10	0.998
GCR	4	1	12	0.984

in hidden layer. The architecture of the developed ANN model is depicted by Fig. 5.7. The value of the correlation coefficient for the developed ANN model is 0.984. Table 5.1 summarizes the important parameters concerning the developed ANN models for the micro-geometry, surface roughness and gear cutting rate.

5.1.3 Comparison of Regression Models and ANN Models

Table 5.2 presents the values of responses predicted by regression and ANN models corresponding to each experimental run. Comparison of regression models and ANN models was done by finding out the average prediction error using Eq. 5.9 for both the models.

$$Prediction\,Error\,(\%) = \frac{Experimental\,value - Predicted\,value}{Experimental\,value} \times 100 \qquad (5.9)$$

It was observed that the ANN models are better predictors of the responses than the regression models. The average prediction error of 3.258 % of ANN models for all responses prove their superiority from RSM models having average prediction error of 8.35 %.

5.2 Optimization of WSEM Process

Optimization of WSEM was done for minimization of micro-geometry errors (F_a and F_p) and surface roughness (R_a and R_{max}), and maximization of gear cutting rate (*GCR*). Single objective Optimization for F_a, F_p, R_a, R_{max} and for *GCR* was done by desirability analysis approach. Multi-objective optimization considering (i) micro-geometry parameters i.e. F_a and F_p simultaneously and (ii) surface roughness parameter i.e. R_a and R_{max} simultaneously was done using RSM-based desirability analysis. While, Multi-objective optimization considering all the five responses simultaneously was done using both RSM-based desirability analysis and BPNN integrated GA approach.

Table 5.2 Regression and ANN predicted values of responses

Expt. No.	Variable input parameters (coded and actual values)								F_a (μm)		F_p (μm)		R_a (μm)		R_{max} (μm)		GCR (mm^3/min)	
	V (Volts)		T_{on} (μs)		T_{off} (μs)		W (m/min)		Regression	ANN	Regression	ANN	Regression	ANN	Regression	ANN	Regression	ANN
1	(1)	15	(0)	0.8	(0)	130	(1)	15	14.17	13.99	28.40	27.05	1.60	1.69	6.93	7.15	36.20	36.92
2	(0)	10	(0)	0.8	(−1)	90	(−1)	9	14.40	14.38	41.38	38.00	1.89	1.98	8.51	9.21	42.65	41.88
3	(−1)	5	(1)	1.0	(0)	130	(0)	12	13.20	13.62	31.90	32.06	1.72	1.80	7.98	8.71	30.31	24.62
4	(0)	10	(0)	0.8	(0)	130	(0)	12	12.89	12.91	14.76	12.20	1.62	1.52	6.45	8.07	30.46	30.94
5	(1)	15	(−1)	0.6	(0)	130	(0)	12	13.59	13.97	21.82	25.67	1.54	1.44	6.55	7.15	30.61	27.40
6	(−1)	5	(0)	0.8	(0)	130	(−1)	9	14.24	14.43	33.70	32.38	1.64	1.68	7.41	8.01	29.30	31.66
7	(−1)	5	(0)	0.8	(1)	170	(0)	12	13.36	12.78	17.30	17.93	1.38	1.34	6.21	6.86	20.14	18.01
8	(0)	10	(1)	1.0	(0)	130	(−1)	9	14.25	14.56	43.20	44.23	1.99	1.81	8.75	8.86	37.42	33.21
9	(1)	15	(0)	0.8	(0)	130	(−1)	9	14.85	15.00	41.20	37.81	1.90	1.68	7.53	8.71	40.56	37.95
10	(0)	10	(0)	0.8	(1)	170	(1)	15	13.93	13.09	20.20	17.43	1.35	1.28	6.06	7.01	28.91	27.62
11	(1)	15	(0)	0.8	(1)	170	(0)	12	16.70	14.47	32.80	34.73	1.63	1.73	6.50	7.28	31.40	28.12
12	(1)	15	(1)	1.0	(0)	130	(0)	12	15.05	15.36	41.40	41.11	1.97	1.92	7.73	8.89	41.57	42.26
13	(0)	10	(0)	0.8	(0)	130	(0)	12	12.89	12.96	14.76	16.52	1.62	1.62	6.45	7.01	30.46	30.93
14	(0)	10	(0)	0.8	(0)	130	(0)	12	12.89	12.66	14.76	12.07	1.62	1.70	6.45	6.79	30.46	33.45
15	(0)	10	(0)	0.8	(0)	130	(0)	12	12.89	13.07	14.76	14.37	1.62	1.59	6.45	6.90	30.46	28.83
16	(1)	15	(0)	0.8	(−1)	90	(0)	12	15.80	15.24	38.27	38.04	1.87	1.87	7.58	8.69	40.78	41.16
17	(0)	10	(0)	0.8	(1)	170	(−1)	9	13.60	13.70	31.15	31.74	1.66	1.68	6.96	7.80	27.21	25.65
18	(0)	10	(1)	1.0	(−1)	90	(0)	12	13.60	13.87	34.50	37.88	1.97	1.98	9.61	9.78	40.63	40.53
19	(−1)	5	(0)	0.8	(−1)	90	(0)	12	14.08	14.28	33.60	33.33	1.61	1.61	7.84	8.20	29.52	28.15
20	(0)	10	(0)	0.8	(0)	130	(0)	12	12.89	13.07	14.76	17.93	1.62	1.52	6.45	7.06	30.46	30.93
21	(0)	10	(−1)	0.6	(−1)	90	(0)	12	14.10	14.31	27.05	34.01	1.53	1.65	7.08	7.43	29.67	28.00
22	(0)	10	(−1)	0.6	(1)	170	(0)	12	13.01	11.78	9.30	8.60	1.29	1.37	6.28	6.51	20.29	25.19
23	(0)	10	(0)	0.8	(−1)	90	(1)	15	12.40	13.66	27.70	25.37	1.59	1.55	7.81	8.24	32.23	32.25
24	(0)	10	(1)	1.0	(0)	130	(1)	15	13.60	13.97	28.20	26.37	1.70	1.73	7.95	8.70	39.04	33.79
25	(−1)	5	(−1)	0.6	(0)	130	(0)	12	13.17	13.26	20.00	20.93	1.27	1.31	6.09	6.66	19.35	17.41
26	(0)	10	(−1)	0.6	(0)	130	(−1)	9	13.30	13.65	22.70	27.34	1.55	1.65	7.11	7.71	26.46	30.65
27	(0)	10	(1)	1.0	(1)	170	(0)	12	14.10	14.49	32.00	32.14	1.72	1.65	7.03	7.89	31.25	28.75
28	(−1)	5	(0)	0.8	(0)	130	(1)	15	12.72	12.66	20.40	16.86	1.34	1.25	6.53	6.40	24.94	21.74
29	(0)	10	(−1)	0.6	(0)	130	(1)	15	10.80	12.24	13.50	14.14	1.26	1.31	6.20	6.63	22.10	21.81

5.2.1 Optimization Using Desirability Analysis

Desirability analysis is a multi-response optimization technique where a single objective function 'D_i' called the 'desirability function' is used for optimization [1]. It is the geometric mean of the individual desirability of all the responses. The parametric combinations that maximize the total desirability are considered to be the optimal parameters. Each response Y_i is converted into an individual desirability function d_i whose value can range from 0 (when the response is outside the acceptable region) to 1 (when the response is at its goal or target value). The more closely the response approaches its goal or target value, the closer the desirability is to 1. Individual desirability d_{ij} is computed for each response j for ith scenario or data set. Desirability of each response can also be assigned a weight w_j according to its relative importance. Equation 5.10 presents the generalized equation of desirability function 'D_i' for the ith data set or scenario [1].

$$D_i = \left[\prod_{j=1}^{n} d_{ij}^{w_j}\right]^{\frac{1}{\sum_{j=1}^{n} w_j}} \tag{5.10}$$

Where n is the number of responses; d_{ij} is the desirability of the jth response for the ith data set or scenario with $0 \leq d_{ij} \leq 1$; and w_j is the weight assigned to jth response.

If T is the target value; L is the acceptable lower limit; U is the acceptable upper limit, and Y_{ij} is the value of the jth response for the ith experimental data, then the values for the individual desirabilities for the different cases can be computed by the following equations:

For the response to maximize

$$d_{ij} = \begin{cases} 0 & Y_{ij} < L \\ \left(\frac{Y_{ij} - L}{T - L}\right)^{w} & L \leq Y_{ij} \leq T \\ 1 & Y_{ij} > T \end{cases} \tag{5.11}$$

For the response to minimize

$$d_{ij} = \begin{cases} 1 & Y_{ij} < T \\ \left(\frac{U - Y_{ij}}{U - T}\right)^{w} & T \leq Y_{ij} \leq U \\ 0 & Y_{ij} > U \end{cases} \tag{5.12}$$

By modifying the Eqs. 5.11 and 5.12, desirability analysis approach was employed to minimize the F_a, F_p, R_a, R_{max} and to maximize the GCR. The following section details the use of desirability approach for the same.

5.2.1.1 Optimization for Micro-geometry

Micro-geometry of gears has two parameters F_a and F_p, and the goal of optimization is to minimize them individually and simultaneously for a desirability value very close to 1. Modifying the Eq. (5.11) to calculate the individual desirabilities for F_a and F_p for the ith data set gives:

$$\left(d_{F_a}\right)_i = \left[\frac{F_{a_{max}} - F_{a_i}}{F_{a_{max}} - F_{a_{min}}}\right]^1 \tag{5.13}$$

$$\left(d_{F_p}\right)_i = \left[\frac{F_{p_{max}} - F_{p_i}}{F_{p_{max}} - F_{p_{min}}}\right]^1 \tag{5.14}$$

Where $F_{a_{min}}, F_{P_{min}}, F_{a_{max}}\ and\ F_{P_{max}}$ are the minimum and maximum values of the $F_a\ and\ F_P$ respectively, based on the experimental data. For minimization of F_a, the overall desirability function for the ith data set can be expressed as:

$$D_i = \left[\left(d_{F_a}\right)_i\right] \tag{5.15}$$

Similarily, for minimization of F_P, the overall desirability function for the ith data set can be expressed as:

$$D_i = \left[\left(d_{F_P}\right)_i\right] \tag{5.16}$$

Single-objective optimization was done for F_a using Eqs. 5.13 and 5.15. The maximum and minimum values chosen from experimental data were 15.2 and 11.7 μm respectively. Table 5.3 presents the optimimum values of WSEM parameters and F_a. Similarily, F_p was optimized using Eqs. 5.14 and 5.16, with maximum value as 44.5 μm and minimum value as 9.4 μm. The corresponding optimum values are given in Table 5.3. In simultaneous optimization, both F_a and F_p were considered equally important, therefore were assigned equal weights of 0.5 each. For minimization of both F_a and F_p simultaneously, the individual desirabilities for the ith data set by modifying the Eq. (5.11) are:

Table 5.3 Optimized values of WSEM parameters and corresponding optimum values of micro-geometry parameters

Objective function	Optimized WSEM parameters				Optimum response	
	V (Volts)	T_{on} (μs)	T_{off} (μs)	W (m/min)	F_a (μm)	F_p (μm)
Min F_a	8.14	0.60	165.4	12.99	11.58	–
Min F_p	8.71	0.64	154.2	12.87	–	8.91
Min F_a *and* F_p	8.62	0.61	159.1	12.98	11.7	9.09

$$(d_{F_a})_i = \left[\frac{F_{a_{max}} - F_{a_i}}{F_{a_{max}} - F_{a_{min}}}\right]^{0.5} \tag{5.17}$$

$$(d_{F_p})_i = \left[\frac{F_{p_{max}} - F_{p_i}}{F_{p_{max}} - F_{p_{min}}}\right]^{0.5} \tag{5.18}$$

The overall desirability function for the *i*th data set can be expressed as:

$$D_i = \left[\{(d_{F_a})_i\}^{0.5}\{(d_{F_P})_i\}^{0.5}\right]^{\frac{1}{0.5+0.5}} \tag{5.19}$$

$$\text{Or} \quad D_i = \left[(d_{F_a})_i(d_{F_P})_i\right]^{0.5} \tag{5.20}$$

Using this concept, Table 5.3 presents the optimum values of the WSEM parameters achieved through desirability analysis for a desirability value of 1.

The validation of the results of optimization for F_a and F_p will be discussed in the section of confirmation experiments and validation of results.

5.2.1.2 Optimization for Surface Roughness

Surface roughness parameters R_a and R_{max} were optimized individually and simultaneously for a desirability value very close to 1. Equations 5.21 and 5.22 give the individual desirabilities for R_a and R_{max} for the *i*th data set.

$$(d_{R_a})_i = \left[\frac{R_{a_{max}} - R_{a_i}}{R_{a_{max}} - R_{a_{min}}}\right]^1 \tag{5.21}$$

$$(d_{R_{max}})_i = \left[\frac{R_{max_{max}} - R_{max_i}}{R_{max_{max}} - R_{max_{min}}}\right]^1 \tag{5.22}$$

Where $R_{a_{min}}, R_{max_{min}}, R_{a_{max}}$ *and* $R_{max_{max}}$ are the minimum and maximum experimental values of the R_a *and* R_{max} respectively. For minimization of R_a, the overall desirability function for the *i*th data set can be expressed as:

$$D_i = \left[(d_{R_a})_i\right] \tag{5.23}$$

Similarily, for minimization of R_{max}, the overall desirability function for the *i*th data set can be expressed as:

$$D_i = \left[(d_{R_{max}})_i\right] \tag{5.24}$$

Using Eqs. 5.21 and 5.23 the optimization was done for R_a. The maximum and minimum values chosen from experimental data were 2 and 1.14 μm respectively.

Table 5.4 Optimized values of WSEM parameters and corresponding optimum values of surface roughness parameters

Objective function	Optimized WSEM parameters				*Optimum response*	
	V (Volts)	T_{on} (μs)	T_{off} (μs)	W (m/min)	R_a (μm)	R_{max} (μm)
Min. R_a	6.31	0.61	164.0	14.47	1.07	–
Min. R_{max}	7.61	0.61	149.9	14.13	–	6.48
Min. R_a *and* R_{max}	6.38	0.63	157.1	14.12	1.08	6.53

Table 5.4 presents the optimimum values of WSEM parameters and R_a. Similarily, optimization was done for R_{max} using Eqs. 5.22 and 5.24, with 9.8 and 6.72 μm being the maximum and minimum values. The results of optimization are given in Table 5.4. Equations 5.25 and 5.26 give the desirabilitities for simultaneous optimization of R_a and R_{max}. Both parameters are equally important and therefore assigned weights of 0.5 each.

$$(d_{R_a})_i = \left[\frac{R_{a_{max}} - R_{a_i}}{R_{a_{max}} - R_{a_{min}}}\right]^{0.5} \tag{5.25}$$

$$(d_{R_{max}})_i = \left[\frac{R_{max_{max}} - R_{max_i}}{R_{max_{max}} - R_{max_{min}}}\right]^{0.5} \tag{5.26}$$

The overall desirability function for the *i*th data set can be expressed as:

$$D_i = \left[\left\{(d_{R_a})_i\right\}^{0.5}\left\{(d_{R_{max}})_i\right\}^{0.5}\right]^{\frac{1}{0.5+0.5}} \tag{5.27}$$

$$\text{Or} \qquad D_i = \left[(d_{R_a})_i (d_{R_{max}})_i\right]^{0.5} \tag{5.28}$$

Using this concept, Table 5.4 presents the optimum values of the WSEM parameters achieved through desirability analysis for a value of 1. The obtained optimum values of WSEM parameters were then standardized based on machine constraints for validation of the results of optimization by confirmation experiments, which will discuss in the section of confirmation experiments and validation.

5.2.1.3 Optimization for Gear Cutting Rate

Optimization of WSEM parameters for maximization of GCR by desirability analysis was done using Eq. 5.12. The overall goal of optimization was to maximize GCR for desirability value of 1. In the present case the desirability function for the *i*th data set can be expressed as:

Table 5.5 Optimized values of WSEM parameters and corresponding GCR

Optimized WSEM parameters				Optimum *GCR* (mm^3/min)
V (Volts)	T_{on} (μs)	T_{off} (μs)	*W* (m/min)	
14.8	0.83	97.1	9.18	42.81

$$(d_{GCR})_i = \left[\frac{GCR_i - GCR_{min}}{GCR_{max} - GCR_{min}}\right] \tag{5.29}$$

Where $GCR_{min}\, and\, GCR_{max}$ are the minimum and maximum values of *GCR* respectively. In this case there is only one response i.e. *GCR* and it was assigned the weight of 1. The overall desirability function for the *i*th data set can be expressed as:

$$D_i = \left[(GCR)_i\right]^1 \tag{5.30}$$

Table 5.5 presents the optimum values of the WSEM parameters and GCR predicted by desirability analysis using Eqs. 5.29 and 5.30. The minimum and maximum values of GCR taken from experimental data are 16.08 and 42.5 mm^3/min respectively. The value of GCR obtained by desirability analysis was validated by confirmation experiments and will be discussed in the next section.

5.2.2 *Multi-objective Optimization*

The conflicting objectives included minimization of micro-geometry errors and surface roughness, and maximization of the process productivity of WSEM (i.e. gear cutting rate). Each response has different set of WSEM parameters for its optimized value. Therefore, a single set of parameters cannot give minimum micro-geometry errors with the best surface finish and highest productivity. A proper trade-off becomes inevitable to satisfy all the above mentioned three machining criteria simultaneously. Multi-objective optimization can be done by assigning the weights to the responses based on their importance. In the present work the task of multi-objective optimization was accomplished by two strategies namely RSM based desirability analysis and BPNN integrated GA approach.

5.2.2.1 Response Surface Methodology Based Desirability Analysis Approach

In the present work, out of the five responses, the most important responses for a miniature gear are errors in total profile and accumulated pitch (F_a, and F_p) followed by the average and maximum surface roughness (R_a and R_{max}) and the GCR.

Table 5.6 Optimized values of WSEM parameters and corresponding values of the responses using RSM based desirability approach

WSEM parameters				Values of the responses				
V (volts)	T_{on} (μs)	T_{off} (μs)	W (m/min)	F_a (μm)	F_p (μm)	R_a (μm)	R_{max} (μm)	GCR (mm^3/min)
9.64	0.64	160.73	14.01	11.77	10.66	1.23	6.60	22

Therefore, weight assigned to F_a, and F_p are 0.3 each, to R_a, and R_{max} are 0.15 each, and to GCR is 0.1. The objective of the present work was to get optimal setting of process parameters that can minimize F_a, F_p, R_a and R_{max}, and maximize GCR simultaneously for the value of desirability function being close to 1. Modified form of the Eq. (5.11) was used to calculate the individual desirabilities for F_a, F_p, R_a, and R_{max}, and modified form of the Eq. (5.12) was used to calculate the individual desirabilities for GCR. The overall desirability function for the ith data set can be computed using the following equation:

$$D_i = \left[\{(d_{F_a})_i\}^{0.3}\{(d_{F_P})_i\}^{0.3}\{(d_{R_a})_i\}^{0.15}\{(d_{R_{max}})_i\}^{0.15}\{(d_{GCR})_i\}^{0.1}\right]^{\frac{1}{0.3+0.3+0.15+0.15+0.1}} \quad (5.31)$$

Maximization of desirabilty function given by Eq. (5.31) was used for multi-objective optmization of WSEM parameters. Table 5.6 presents the optimum values of the WSEM parameters achieved through desirability analysis and corresponding predicted values of the responses.

5.2.2.2 Back Propagation Neural Network Integrated Genetic Algorithm Approach

Genetic algorithms (GA) are very efficient evolutionary optimization technique mimicking the natural evolution process using reproduction, crossover and mutation operations based on principles of genetics [7–9]. GA starts and subsequently works with a set of solutions, known as population, and tries to reach towards the global optimum solution in a defined number of generations. Consequently, they are very effective for finding optimum solutions of the unconstrained or constrained single or multi-objective optimization problems particularly those involving implicit dependence of objective functions on the variables to be optimized. Each individual element in the population represents a combination of feasible values (as defined by the variable bounds) of all the variables and is referred as chromosome. It contains a string of binary numbers (in case of binary coded GA) or real numbers (in case of real coded GA) known as genes. The GA discretizes the entire search space as defined by the variable bounds, and this discrete value depends on the string length. During the optimization process, the population of current chromosomes (or solutions) is replaced by a new generation of chromosomes by three operations of *crossover*, *mutation*, and *reproduction*. The new population may contain both parent chromosomes and newly generated chromosomes called *offsprings*. The two operators namely *crossover* and *mutation* are used to generate the offspring chromosomes. The crossover involves crossing of two parent strings at the defined site to form two new *offsprings*. The resultant offsprings may be better or worse than their parents. Mutation refers to a process of flipping a gene at the defined site of a selected chromosome and is responsible for bringing diversity in the population. A fitness function representing the objective function is used to evaluate the population of chromosomes for each generation. The chromosomes with higher fitness values are selected for the next generation and this is referred as *reproduction*. This cycle of three operators is repeated either the attainment of the optimum solution or satisfaction of the stopping criteria (i.e. defined number of generations or no further improvement in the optimum solution between the successive generations).

In the present work, GA were integrated with BPNN for multi-objective optimization of the conflicting objectives of micro-geometry, surface roughness and productivity of miniature gears by WSEM. Figure 5.8 shows the various tasks performed in this approach with the help of a flowchart.

First, a BPNN models for all five responses were developed by training the NN using their experimental data. The developed NN model has 12 neurons in hidden

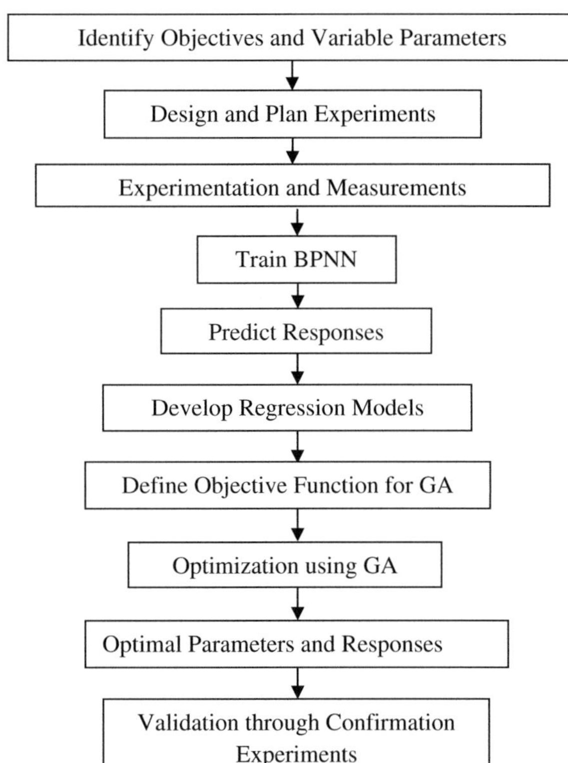

Fig. 5.8 Tasks in multi-objective optimization using BPNN integrated GA

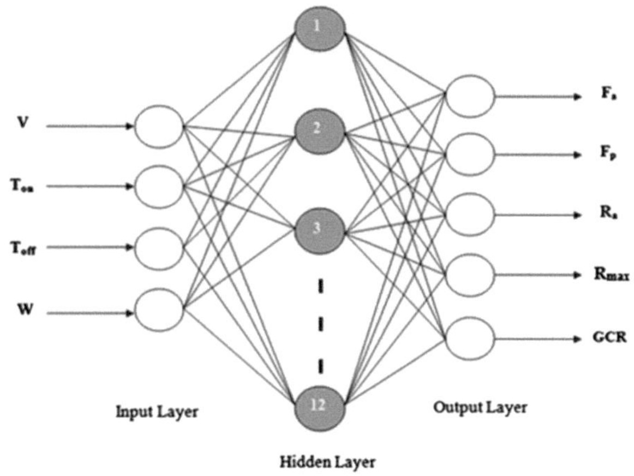

Fig. 5.9 Architecture of the developed BPNN model for multi-objective optimization

layer and the value of regression coefficient was 0.9909. Its architecture is given in Fig. 5.9. Then BPNN predicted values of the responses were used to develop the mathematical models for them in terms of the WSEM parameters by regression analysis. Equations 5.32–5.36 present these models in terms of only significant parameters and interactions which were identified through ANOVA of the BPNN predicted results. These equations were used to form the objective function for multi-objective optimization by GA.

$$\begin{aligned} \boldsymbol{F_a} = {} & 49.57 - V - 28.82T_{on} - 0.17T_{off} - 1.49W + 0.032V^2 + 7.57T_{on}^2 \\ & + 0.00026T_{off}^2 + 0.04W^2 + 0.26VT_{on} + 0.1T_{on}T_{off} \end{aligned} \quad (5.32)$$

$$\begin{aligned} \boldsymbol{F_p} = {} & 418.9 - 9.7V - 304.21T_{on} - 1.93T_{off} - 17.3W + 0.34V^2 + 169.66T_{on}^2 \\ & + 0.004T_{off}^2 + 0.67W^2 + 0.015VT_{off} + 0.61T_{on}T_{off} \end{aligned} \quad (5.33)$$

$$\begin{aligned} \boldsymbol{R_a} = {} & 5.64 + 0.15V - 1.77T_{on} - 0.009T_{off} - 0.54W - 0.003V^2 - 0.0005VT_{off} \\ & + 0.221T_{on}W + 0.0007T_{off}W \end{aligned} \quad (5.34)$$

$$\begin{aligned} \boldsymbol{R_{max}} = {} & 23.46 - 9.1T_{on} - 0.07T_{off} - 1.3W + 12.68T_{on}^2 + 0.0003T_{off}^2 + 0.04W^2 \\ & - 0.04T_{on}T_{off} \end{aligned} \quad (5.35)$$

$$\begin{aligned} \boldsymbol{GCR} = {} & 148.91 - 1.71V + 21.37T_{on} - 0.311T_{off} - 17.27W - 0.07V^2 + 0.29W^2 \\ & + 0.16VW - 0.29T_{on}T_{off} + 4.96T_{on}W + 0.03T_{off}W \end{aligned} \quad (5.36)$$

In the present work, the binary coded GA were used for multi-objective optimization of WSEM parameters using an objective function (Eq. 5.37) which is the weighted sum of the five individual objectives.

$$Min\, F = \left[w_1 f(F_a) + w_2 f(F_p) + w_3 f(R_a) + w_4 f(R_t) - w_5 f(MRR)\right] \quad (5.37)$$

The variable bounds defined by:

$$5 \leq \mathrm{V} \leq 15\ (\text{volts})$$

$$0.6 \leq \mathrm{T_{on}} \leq 1\ (\mu\text{s})$$

$$90 \leq \mathrm{T_{off}} \leq 170\ (\mu\text{s})$$

$$9 \leq \mathrm{W} \leq 15\ (\text{m/min}).$$

The five objectives involved in the Eq. (5.37) include:

Objective 1 Minimize profile error, $F_a = f_1\left(V, T_{on}, T_{off}, W\right)$
Objective 2 Minimize pitch error, $F_p = f_2\left(V, T_{on}, T_{off}, W\right)$
Objective 3 Minimize average roughness, $R_a = f_3\left(V, T_{on}, T_{off}, W\right)$
Objective 4 Minimize maximum roughness, $R_{max} = f_4\left(V, T_{on}, T_{off}, W\right)$
Objective 5 Maximize $GCR = f_5\left(V, T_{on}, T_{off}, W\right)$

These objectives were evaluated using their regression equations (i.e. Eqs. 5.32–5.36). Values of the weights assigned to them are same (i.e. w_1 = 0.3; w_2 = 0.3; w_3 = 0.15; w_4 = 0.15; and w_5 = 0.1) as used in multi-objective optimization by

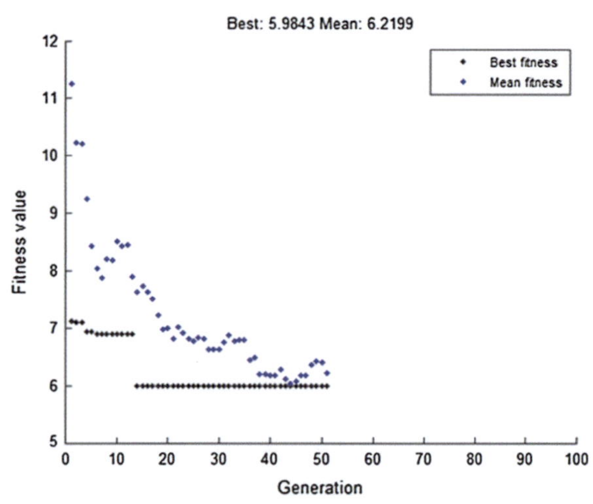

Fig. 5.10 Performance of the GA used with the number of generations

Table 5.7 Optimized values of WSEM parameters and corresponding values of the responses using BPNN integrated GA approach

WSEM parameters				Values of the responses				
V (volts)	T_{on} (μs)	T_{off} (μs)	W (m/min)	F_a (μm)	F_p (μm)	R_a (μm)	R_{max} (μm)	GCR (mm^3/min)
10.13	0.64	169.9	12.92	11.91	8.86	1.28	6.73	19.3

RSM based desirability analysis. The GA parameters used for optimization are: *Population size* = 50; *Reproduction operator*: Roulette wheel; *Type of crossover operator and its probability*: Two point with 0.8; *Type of mutation operator and its probability*: Uniform with 0.05. Figure 5.10 illustrates the performance of the GA used with the number of generations. It is clear from this figure that the mean value of the fitness reaches to the best fitness value in forty five generations and that it does not change significant after 50 generations. This indicates that the WSEM parameters have attained their optimum values as mentioned in the Table 5.7.

5.3 Experimental Validation of Optimization Results

The optimum values of WSEM parameters obtained for each case discussed in the previous section were then standardized based on machine constraints. Using the standardized optimum values, total 9 confirmation experiments were conducted to validate the optimum values of the WSEM parameters and corresponding responses.

5.3.1 For Micro-geometry

A total of 3 experiments were conducted included one each for individual optimization of F_a and F_p, and one for simultaneous optimization of them. Table 5.8 presents the optimum values of F_a, F_p and F_a–F_p respecively obtained by confirmation experiments conducted at the standardized values of WSEM parameters and given by desirability analysis.

Very close agreement between the experimenral results and those obtained by the desirablity analysis validates the optimization results. As given in Table 5.10, the values for multi-objective optimization of F_a (11.5 μm) and F_p (9.1 μm) obtained in confirmation experiments categorize the gear in quality no. DIN 7 and DIN 5 respectively. It was also observed that the optimum values of F_a and F_p predicted by ANN are closer to the experimental results than the values predicted by desirability analysis.

Table 5.8 Value of WSEM parameters and corresponding values of F_a, F_p, and F_a–F_p obtained in confirmation experiments and predicted by desirability analysis

Response	WSEM parameters								Desirability analysis	Confirmation experiment
	Optimized				Standardized					
	V (volts)	T_{on} (µs)	T_{off} (µs)	W (m/min)	V (volts)	T_{on} (µs)	T_{off} (µs)	W (m/min)		
F_a (µm)	8.14	0.60	165.4	12.99	8	0.6	165	13	11.58	11.1
F_p (µm)	8.71	0.64	154.2	12.87	9	0.65	155	13	8.9	8.4
F_a (µm)	8.62	0.61	159.1	12.98	9	0.6	160	13	11.7	11.5
F_p (µm)									9.09	9.1

Table 5.9 Value of WSEM parameters and corresponding value of R_a, R_{max}, and R_a–R_{max} obtained in the confirmation experiment and predicted by desirability analysis

Response	WSEM parameters								Desirability analysis	Confirmation experiment
	Optimized				Standardized					
	V (volts)	T_{on} (µs)	T_{off} (µs)	W (m/min)	V (volts)	T_{on} (µs)	T_{off} (µs)	W (m/min)		
R_a (µm)	6.31	0.61	164.0	14.47	6	0.6	165	14	1.07	1.05
R_{max} (µm)	7.61	0.61	149.9	14.13	8	0.6	150	14	6.48	6.34
R_a (µm)	6.38	0.61	157.1	14.12	6	0.6	160	14	1.04	1.10
R_{max} (µm)									6.53	6.39

5.3.2 For Surface Roughness

Table 5.9 presents the optimum values of WSEM parameters and corresponding values of R_a, R_{max} and R_a–R_{max} obtained by confirmation experiments at standardized WSEM parameters and by desirability analysis.

It is evident that the values of R_a and R_{max} obtained at confirmation experiments are very close to the values given by desirability analysis, which validates the optimization results. It was also observed that ANN predicted values of R_a and R_{max} are closer to the experimental values than desirability predicted values.

5.3.3 For Gear Cutting Rate

Very close agreement between experimentally obtained (42.97 mm^3/min) and desirability predicted (42.81 mm^3/min) values of GCR as given in Table 5.10 validates the results of optimization. It was observed that ANN predicted results are much closer to the experimental results.

5.3.4 For Multi-objective Optimization

Table 5.11 presents the optimized and standardized WSEM parameters, and corresponding values of the responses for RSM based desirability approach and BPNN integrated GA approach. Very close agreements between the values of the responses given by approaches and confirmation experiments validate the results of optimization. The optimum values of F_a, F_p and R_a given by BPNN-GA technique are better than those given by RSM-desirability approach. But, the values of R_{max} and *GCR* by RSM-desirability approach are better than those given by BPNN-GA.

It was observed that optimum values of F_a–F_p, R_a–R_{max} and GCR need machining at different set of parameters. By assigning proper weights to the responses based on their importance multi-objective optimization was done. The miniature gear manufactured at combination of 10 V, 0.65 µs T_{on}, 170 µs T_{off} and 13 m/min W given by BPNN-GA approach, had quality no. DIN-7 and DIN 5 for F_a and F_p respectively and R_a value 1.26 µm.

Table 5.10 Values of *GCR* obtained in confirmation experiments and given by desirability analysis

Response	WSEM parameters								Values of GCR	
	Optimized				Standardized					
	V (volts)	T_{on} (μs)	T_{off} (μs)	*W* (m/min)	*V* (volts)	T_{on} (μs)	T_{off} (μs)	*W* (m/min)	Desirability analysis	Confirmation experiment
GCR (mm^3/min)	14.8	0.83	97.1	9.18	15	0.85	100	9	42.81	42.97

Table 5.11 Values of WSEM parameters and corresponding values of the responses for multi-objective optimization by RSM based desirability and BPNN integrated GA

Approach	WSEM parameters				Values of the responses				
	V (volts)	T_{on} (μs)	T_{off} (μs)	W (m/min)	F_a (μm)	F_p (μm)	R_a (μm)	R_{max} (μm)	*GCR* (mm^3/min)
RSM based desirability analysis	9.64	0.64	160.7	14.01	11.77	10.66	1.23	6.60	22
Confirmation experiments	10	0.65	160	14	11.95	10.2	1.29	6.64	20.83
BPNN integrated GA approach	10.13	0.64	169.9	12.92	11.91	8.86	1.28	6.73	19.30
Confirmation experiments	10	0.65	170	13	11.9	8.6	1.26	6.67	20.08

5.4 Quality Aspects of Miniature Gears Manufactured at Optimized WSEM Parameters

Quality aspects namely micro-geometry, macro-geometry, bearing area curve and surface integrity of the miniature gears manufactured at different sets of optimum WSEM parameters are given in this section.

5.4.1 *Micro-geometry*

Figure 5.11 presents the results of micro-geometry of the gear manufactured at optimized (using Desirability analysis) WSEM parameters in the confirmation experiment conducted for simultaneous minimization of both F_a and F_p. This gear had *DIN* quality number 7 for profile (with F_a value of 11.5 μm) and *DIN* quality number 5 for pitch (with F_p value of 9.1 μm), which are much superior to the miniature gears manufactured by conventional processes.

Figure 5.12 presents the results of F_a and F_p for the miniature gear manufactured at optimum WSEM parameters corresponding to the multi-objective optimization done using BPNN-GA technique. This gear also had DIN-7 for F_a (11.9 μm) and DIN-5 for F_p (8.6 μm). The quality of these gears is equivalent to the quality produced by highly precise gear finishing processes such as gear grinding, shaving and honing (refer Table 1.1 of Chap. 1).

(a)

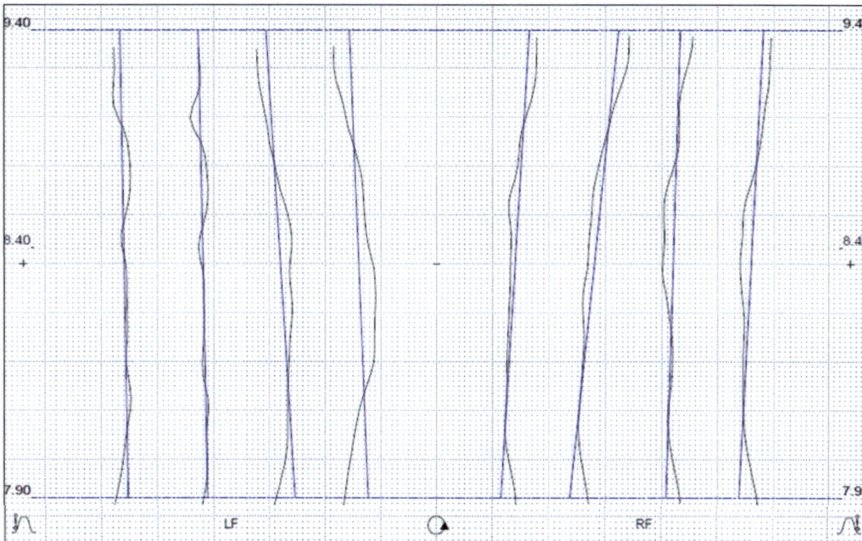

[...]	x	#10	#7	#4	#1	Tooth	#1	#4	#7	#10	x	[...]
-/-	10.2	6.5	6.5	12.7	15.0	Fa	11.1	18.1	11.0	10.9	12.8	-/-
-/-	9.2	6.6	5.8	10.8	13.6	ffa	8.4	8.9	8.6	9.9	8.9	-/-
-/-	5.9	3.0	3.4	10.4	6.8	fHa	10.1	17.4	5.1	8.6	10.3	-/-

(b)

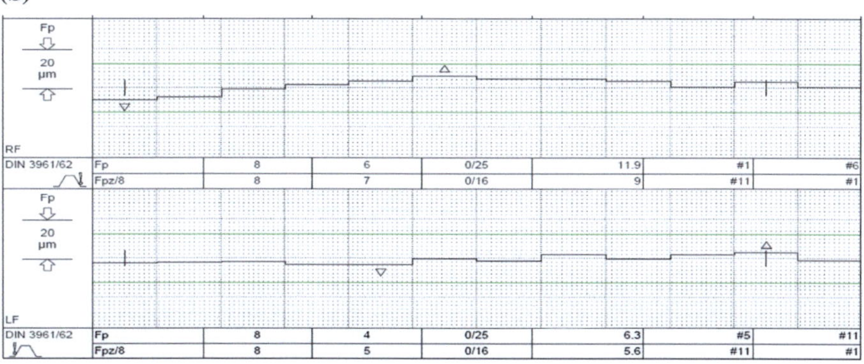

DIN 3961/62						
Fp	8	6	0/25	11.9	#1	#6
Fpz/8	8	7	0/16	9	#11	#1

DIN 3961/62						
Fp	8	4	0/25	6.3	#5	#11
Fpz/8	8	5	0/16	5.6	#11	#1

Fig. 5.11 Results of **a** total profile error **b** accumulated pitch error for *right flanks* (RF) and *left flanks* (LF) of the miniature gear manufactured in confirmation experiment at WSEM parameters for simultaneous optimization of F_a and F_p [10]

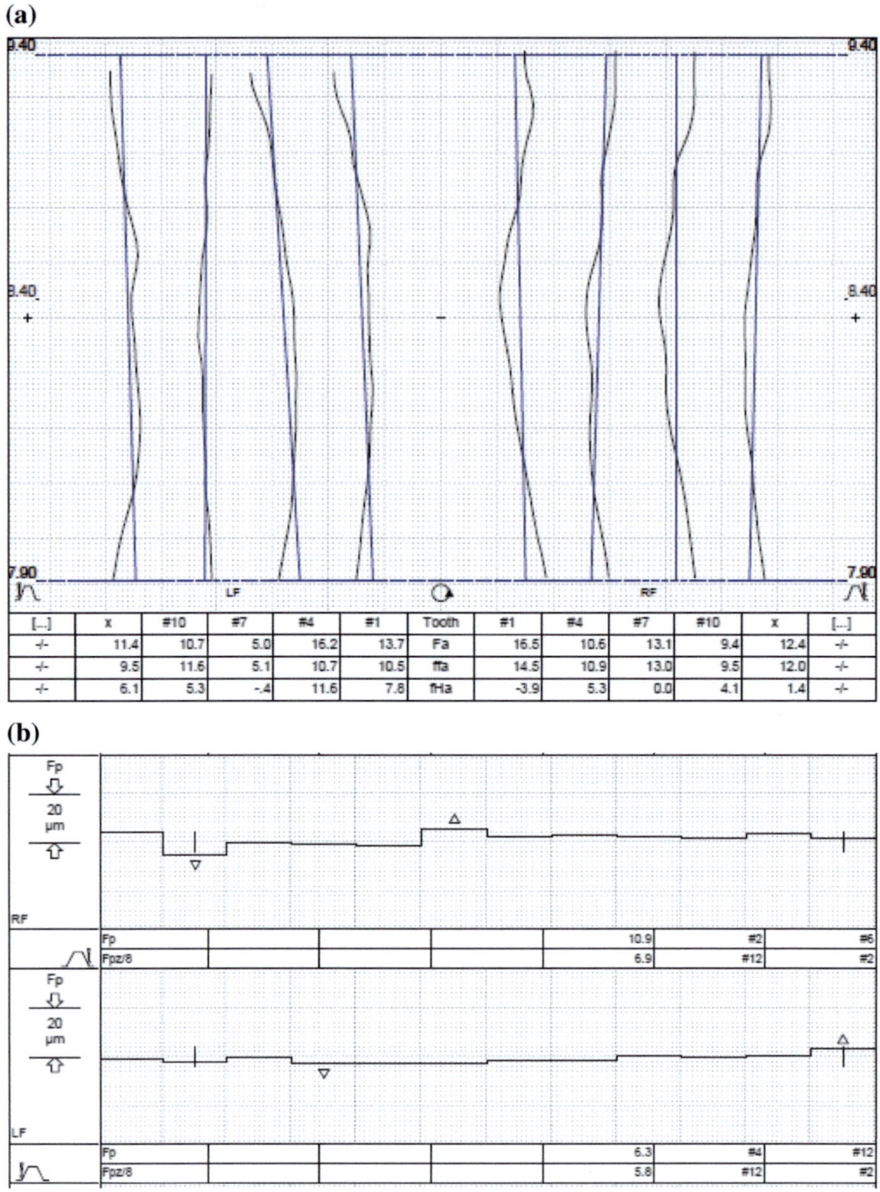

Fig. 5.12 Results of **a** total profile error **b** accumulated pitch error for *right flanks* (RF) and *left flanks* (LF) of the miniature gear manufactured in confirmation experiment at optimum WSEM parameters given by BPPN-GA

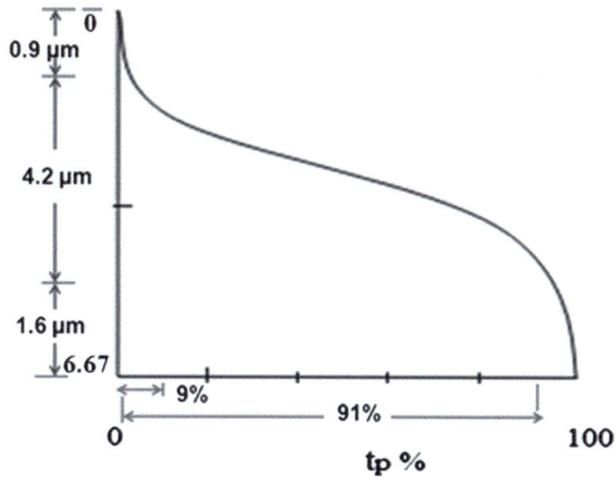

Fig. 5.13 BAC curve for the miniature gear manufactured at optimum WSEM parameters given by BPNN-GA

5.4.2 *Macro-geometry*

The gear manufactured for optimum values of F_a and F_p had very low values of macro-geometry errors i.e. deviation in span (3.6 μm), deviation in chordal tooth thickness (3.2 μm) and deviation in the outside diameter (8.4 μm). Whereas the gear manufactured for BPNN-GA based multi-objective optimization had 4.2 μm span deviation, 4.6 μm tooth thickness deviation and 9.5 μm deviation in outside diameter.

5.4.3 *Bearing Area Curve*

Figure 5.13 illustrates a BAC curve for the miniature gear manufactured at WSEM parameters optimized by BPNN-GA approach during multi-objective optimization. The value of maximum roughness for this gear is 6.67 μm. The values of R_{pk} and M_{r1} are 0.9 μm and 9.1 % which indicates that about 9.1 % of initial surface profile area is likely to wear out during initial run-in period with surface peaks reduced by of 0.9 μm depth. The value of R_k is about 4.2 μm and corresponding bearing length is 70 % i.e. more than 70 % of the surface roughness data are confined within 4 μm depth. M*aterial component corresponding to the lower limit of the roughness core* 'M_{r2}' is about 91 % and *reduced valley depth* 'R_{vk}' is about 1.6 μm. Although, the maximum roughness value for this gear is high as compared to the best gear manufactured in pilot experiment but it possesses comparatively better bearing area characteristics.

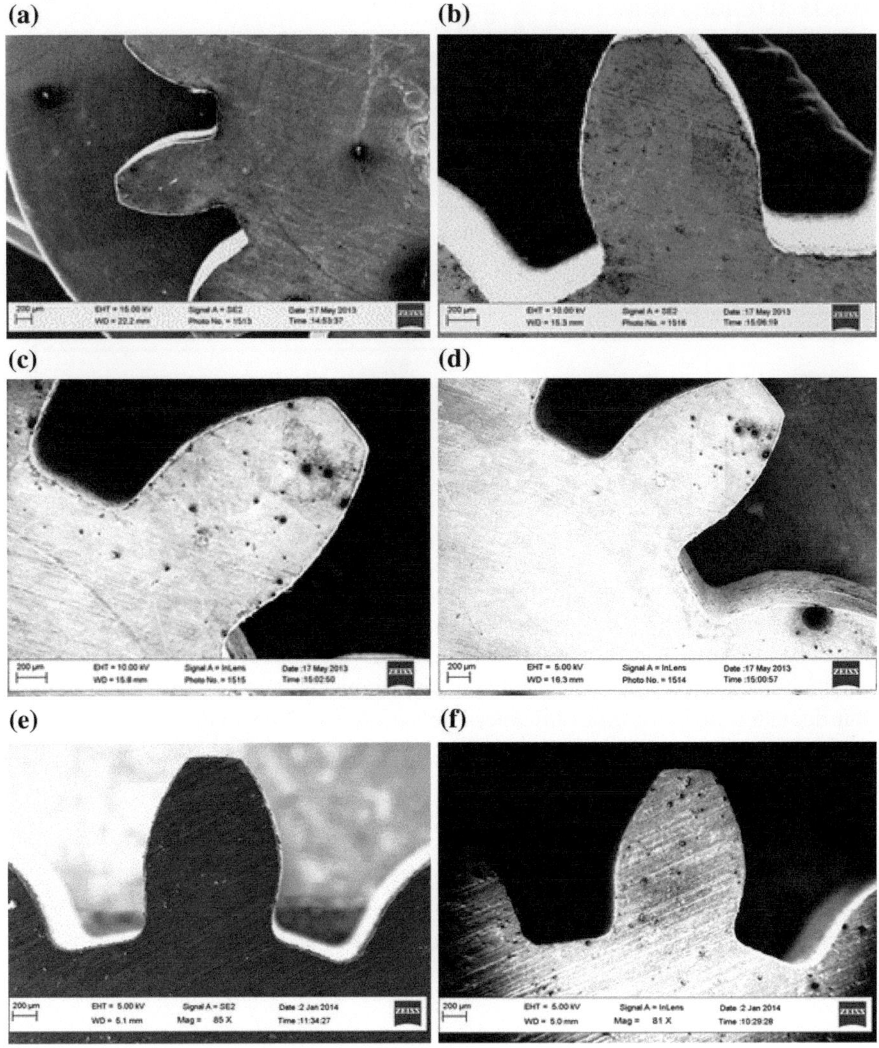

Fig. 5.14 Micrographs of WSEMed miniature gears [10–12]. **a** at 25 X, **b** at 60 X, **c** at 80 X, **d** at 40 X, **e** at 85 X, **f** at 80 X

5.4.4 Subsurface Integrity

A micro graphical study of the miniature gears (Fig. 5.14) machined for *WSEM* parameters optimized for: surface roughness (Fig. 5.14a, b), micro-geometric accuracy (Fig. 5.14c, d) and for the multi-objective optimized set (Fig. 5.14e, f) reveals the formation of burr-free uniform teeth profiles without any undercutting at the root area.

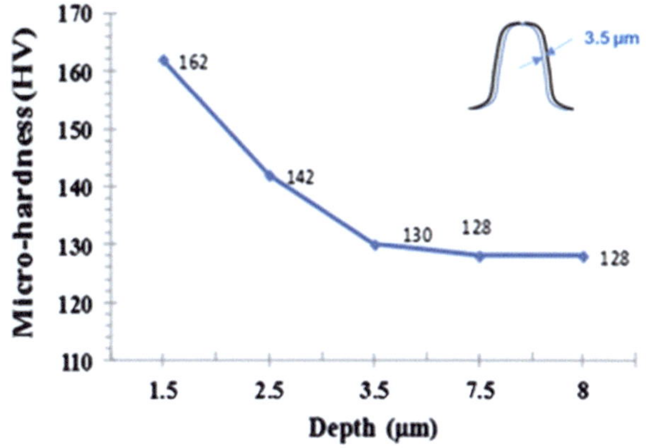

Fig. 5.15 Results of micro-hardnes measurements for the miniature gear manufactured at WSEM parameters optimized by BPNN-GA

Micro-hardness measurement of the WSEMed miniature gear produced at the multi-objectively optimized (using BPNN-GA) parameters displayed the presence of thin i.e. 3.5 μm recast layers (Fig. 5.15) that seems not to have adverse thermal effects on gear teeth surfaces. This is largely the result of the high thermal conductivity of the gear material (brass) that is conducive to rapid dissipation of heat during machining. This represents the capability of WSEM to manufacture miniature gears with almost negligible thickness of the recast layer.

References

1. Montgomery DC (2001) Design and analysis of experiments. Wiley, New York
2. Tzeng C, Yang Y, Hsieh M, Jeng MC (2011) Optimization of wire electrical discharge machining of pure tungsten using neural network and response surface methodology. Proc IMechE, Part-B J Eng Manuf 225:841–852
3. Cai C (2012) Application of neural network in the process modeling of WSEM. Commun Comp Inf Sci 288:285–293
4. Chen H, Lin J, Yang Y, Tsai C (2010) Optimization of wire Spark Erosion machining for pure tungsten using a neural network integrated simulated annealing approach. Expt Syst Appl 37 (10):7147–7153
5. Fausett L (2006) Fundamentals of neural networks. Pearson Education, New Delhi
6. Huang S, Huang Y (1991) Bound on the number of hidden neurons in multi-layer perceptrons. IEEE Trans Neural Netw 2(1):47–55
7. Deb K (2001) Multi-objective optimization using evolutionary algorithms. Willey, Chichester
8. Gen M (1997) Genetic algorithm and engineering design. Wiley, New York
9. Goldberg D (1989) Genetic algorithms in search, optimization and machine learning. Addison Wesley Longman Publishing Co., Inc., MA
10. Gupta K, Jain NK (2014) Analysis and optimization of micro-geometry of miniature gears manufactured by wire electric discharge machining. Precis Eng 38(4):728–737
11. Gupta K, Jain NK (2014) Analysis and optimization of surface finish of wire electrical discharge machined miniature gears. Proc IMechE B J Eng Manuf 228(5):673–681
12. Gupta K, Jain NK (2014) Comparative study of wire-EDM and hobbing for manufacturing high quality miniature gears. Mater Manuf Processes 29:1470–1476

Chapter 6
Comparative Study, Conclusions and Future Avenues

6.1 Comparison Between Wire Spark Erosion Machining and Hobbing

Gear Hobbing is a material removal (machining) type gear manufacturing process which is most extensively used among all conventional processes to manufacture miniature gears. As discussed in Chap. 1, hobbing suffers from following major limitations: (i) generation of undesirable cutting tool marks on gear tooth surface; (ii) inability to manufacture internal and conical gears; (iii) inability to perform tooth modifications such as tip relief, root alteration, profile correction, and crowning; (iv) special tooling requirements; (v) long setup times; (vi) uneconomical for small volume production or prototype gears; and (vii) poor quality of gears (DIN-9) thus necessitating subsequent finishing operation such as shaving, grinding, honing, burnishing etc. to produce a fit for purpose quality [1, 2].

The subsequent finishing operations themselves require large amounts of resources and energy, which adds extra cost and adversely affects the environment. Main reasons responsible for errors in the macro and micro-geometry of gears manufactured by hobbing are (i) incorrect sharpening of the hob cutter; (ii) backlash and runout in the hob cutter and gear blank spindle system; (iii) inaccurate alignment of the gear blank; and (iv) vibration and chatter due to the worn-out cutting edges of hob cutter [1–3].

Table 6.1 presents comparison of the quality parameters achieved for the best quality WSEMed miniature gear (multi-objectively optimized) with the best quality hobbed gear (selected from the number of hobbed gears manufactured) made of same material and design specifications. Following sections describe the detailed comparison between WSEM and hobbing for manufacturing of miniature gears of same specifications.

© Springer Science+Business Media Singapore 2017
K. Gupta and N.K. Jain, *Near-Net Shape Manufacturing of Miniature Spur Gears by Wire Spark Erosion Machining*, Materials Forming, Machining and Tribology, DOI 10.1007/978-981-10-1563-2_6

Table 6.1 Comparison of WSEMed gear and hobbed gear on the basis of micro-geometry, macro-geometry and surface finish [2]

Micro-geometry parameter (μm)	WSEMed gear		Hobbed gear	
	Value	DIN No.	Value	DIN No.
Profile error	11.9	7	30.2	10
Pitch error	8.6	5	48	10
Surface roughness parameter	*Value*			
Average roughness (μm)	1.26		0.2	
Maximum roughness (μm)	6.67		1	
Skewness	−0.165		−0.360	
Kurtosis	2.455		2.635	
Macro-geometry parameter (μm)	*Value*			
Deviation in span	4		44	
Deviation in tooth thickness	5		47	
Deviation in outside diameter	10		97	

6.1.1 Micro-geometry and Macro-geometry

It has been discussed in Chap. 1 that profile error and pitch error are relatively important micro-geometry parameters for the miniature gears made of brass (primarily used as motion transmission gears) because their values determine the noise behavior and motion transfer characteristics, respectively. As given in Table 6.1, the measured values of profile error and pitch error for the best quality WSEMed gear are 11.9 μm and 8.6 μm respectively. This assigns it DIN-7 for profile and DIN-5 for pitch. Whereas the best quality hobbed gear has DIN standard 10 for profile and pitch both with measured values of profile error as 30.2 μm and pitch error as 48 μm. Therefore, it can be concluded that manufacturing quality of WSEMed miniature gear is much more superior to the hobbed gear and even at par with the quality given by most of the precision finishing processes which meets the requirements of various precision applications. It is also evident from Table 6.1 that deviations in the macro-geometry parameters (i.e. span, tooth thickness, and outside diameter) for the hobbed gear are also very high as compared to that for the WSEMed gear.

6.1.2 Surface Roughness and Flank Surface Topography

Average roughness, maximum roughness, skewness and kurtosis are important surface roughness parameters which affect service life, wear behavior and govern the tribology characteristics of the miniature gears. On the basis of measured average and maximum surface roughness, skewness and kurtosis, as mentioned in Table 6.1, the hobbed gear was found to have better surface finish in terms of lower values of average and maximum roughness, and skewness than WSEMed gear. While the difference between the values of kurtosis for WSEMed and hobbed gear was insignificant. But, the surface roughness values obtained for WSEMed gear is acceptable for miniaturized applications [4–8].

Despite the hobbed gear having better surface finish some tool (hob cutter) marks were observed on its tooth flank surfaces while examining its topography. Figure 6.1 depicts the flank surface topography of the miniature gears manufactured by WSEM and hobbing. The entire flank surface was divided into nine equally spaced sections along the face width and four sections along the profile for a randomly chosen gear tooth for both the gears. The blue grid lines represent the theoretical flank surfaces, and the black grid lines represent the actual flank surfaces. It can be seen that the deviations in actual flank surface from the theoretical surface are more for hobbed gear (Fig. 6.1b) than that for the WSEMed gear (Fig. 6.1a). The tool marks on the teeth surface of the hobbed gear are the main reason responsible for that. On the basis of topography study, it can be concluded that despite having a superior surface finish, the hobbed gear cannot guarantee better performance during its service life until the tool marks are removed by a subsequent finishing operation. The WSEMed gear has very little deviation of actual flank surface from the theoretical one i.e. no marks were observed on flank surfaces. This ensures better flank surface condition of WSEMed gear in comparison to the hobbed gear.

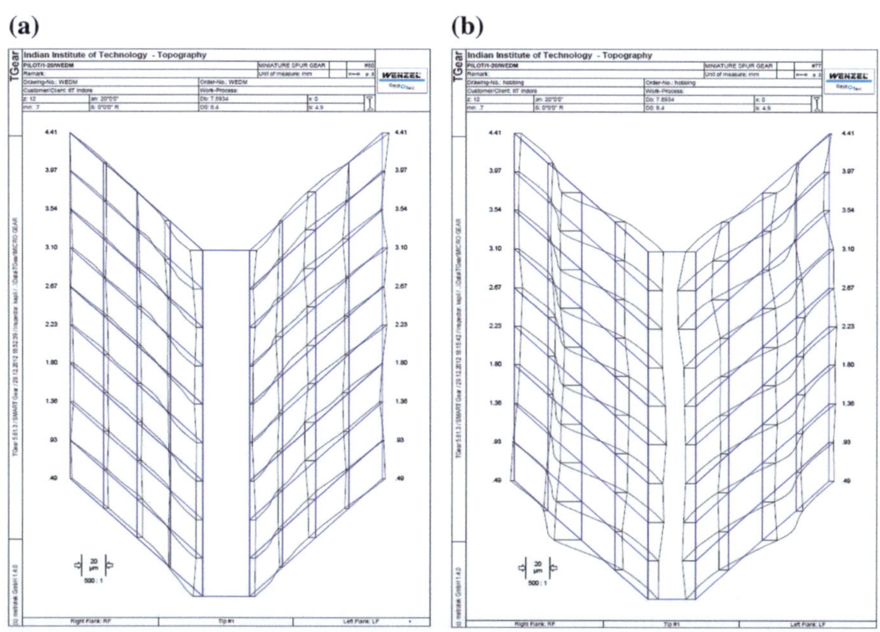

Fig. 6.1 Flank surface topography of **a** WSEMed gear, and **b** hobbed gear [2]

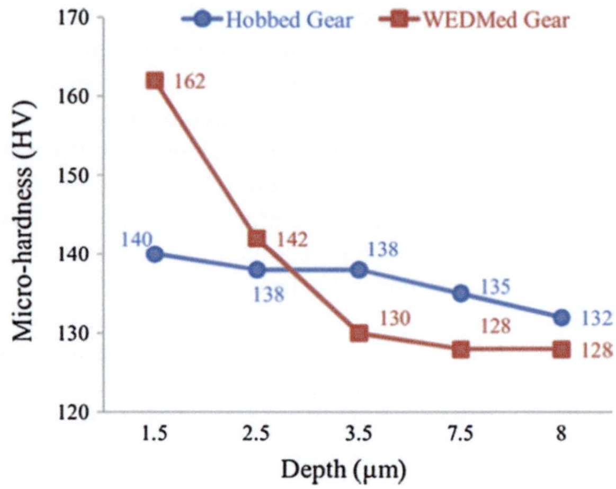

Fig. 6.2 Variation in micro-hardness of miniature gears manufactured by WSEM and hobbing [2]

6.1.3 *Micro-hardness*

Micro-hardness evaluation can be used to find presence of recast layer or surface hardening along the gear tooth surface. Figure 6.2 depicts the variation in micro-hardness of the tooth surfaces of miniature gear manufactured by WSEM and hobbing. It is evident from this figure that micro-hardness of WSEMed gear is high just at the few microns below the flank surface, decreasing up to the depth of 3.5 μm and after that remains almost constant. This variation indicates existence of a recast layer of an approximate thickness of 3.5 μm along the surface of the WSEMed gear tooth. Such a small value of the thickness of recast layer indicates no adverse thermal effects of WSEM process on miniature gear, as high thermal conductivity of brass led to rapid dissipation of heat during WSEM process. Micro-hardness evaluation of the hobbed gear also does not reveal any significant change in it and presence of any recast layer along the depth of the tooth surface.

6.1.4 *Micrographs*

Micrographs of the tooth profile of miniature gear are the useful indicators of substantial tooth form error, burr formation and presence of any other surface defects. Figures 6.3 and 6.4 show scanned electron micrographs of miniature gears manufactured by WSEM and hobbing respectively. Higher amount of profile error along with undercutting at root area can be seen on the hobbed gear (see Fig. 6.4a–e) in comparison to the WSEMed gear which exhibits uniform, burr-free tooth profile with no undercutting at root area (see Fig. 6.3a–e). WSEMed gear tooth surface was also observed to be free from cracks, voids, and globules etc. This proves the superiority of WSEM process to the hobbing process.

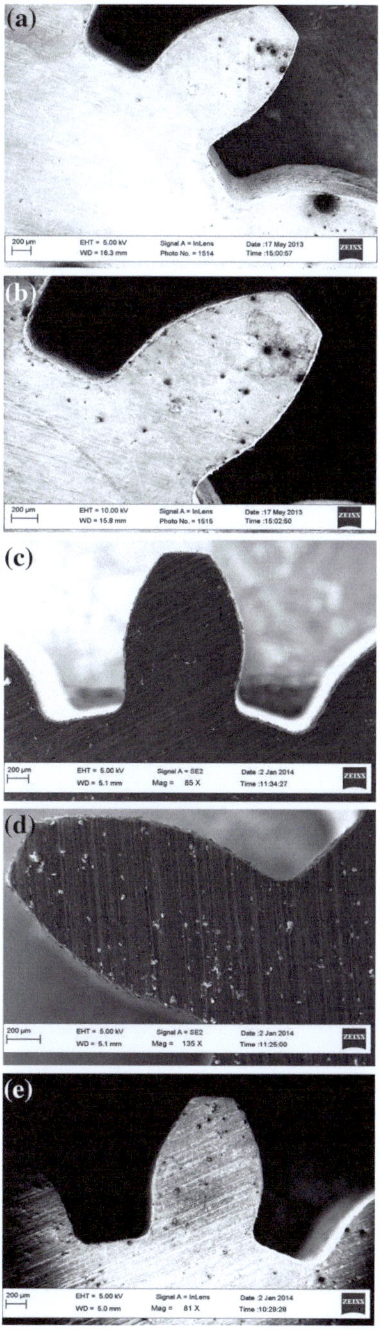

Fig. 6.3 Micrographs of WSEMed gear [2]

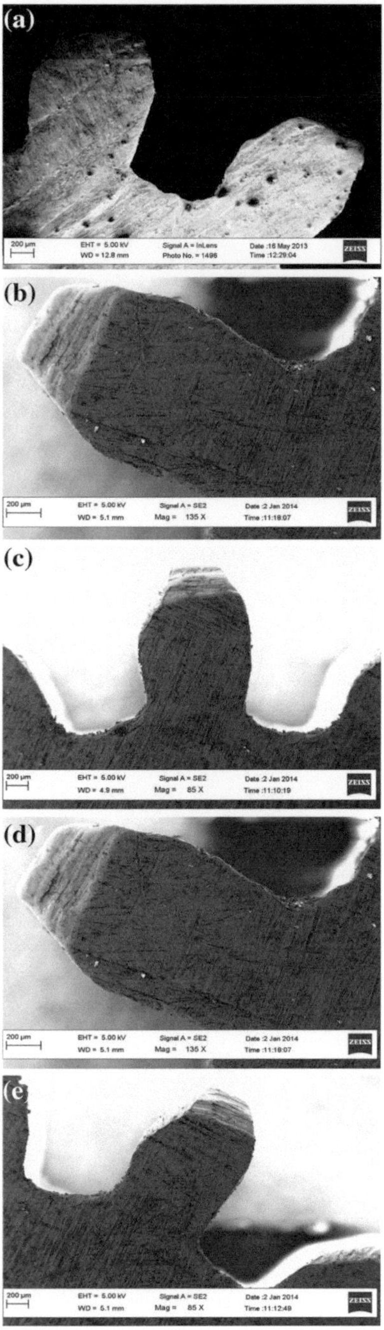

Fig. 6.4 Micrographs of hobbed gear [2]

Since, WSEMed gear is found to have better micro and macro-geometry characteristics, better flank topography and uniform tooth profile, therefore, it can be concluded that the WSEM process is superior to the hobbing process for manufacturing of high quality miniature gears. It can also be extended to manufacture micro size gears by using the wire of comparatively low diameter. Since, surface finish of WSEMed gear is not good, consequently more research is required to improve the finish for longer service life of the gear.

6.1.5 Comparison Based on Other Capabilities

Competitive and sustainable manufacturing requires reduction in the number of manufacturing stages i.e. short process chain, minimum wastages, and energy and resource efficiency without compromising the functionality and quality of the products [6]. WSEM promises to be such process because (i) it does not require any subsequent finishing processes unlike hobbing; (ii) it does not require hydrocarbon-based harmful cutting fluids which gear hobbing needs; (iii) it does not demand special cutting tools and processes as required by gear hobbing; (iv) it can modify the tooth profile, including tip relief, root alteration, and crowning; (v) it can manufacture the internal or external gears of any shape with accuracy equivalent to the gear grinding; (vi) it can manufacture unique shapes such as partial gears, non-involute profile, and noncircular gears; (vii) it requires lower setup time and overall process time, making the gear prototype manufacturing and low-volume production cheaper; (viii) material wastage in WSEM is very less because manufacturing of gears by WSEM is analogous to a punching process. Moreover, if process is controlled precisely then simultaneous manufacturing of external and internal gear is possible; (ix) the cost of manufacturing a fine-pitched gear of DIN-5 quality by WSEM is significantly lower to hobbing and conventional set of processes; (x) The overall productivity of WSEM is higher for manufacturing of high-quality miniature gears [2, 6].

In essence, potential single stage manufacturing of gears by wire electric-discharge machining (WSEM) ensures higher quality; energy, resource and economic efficiency; and better ecological prospects.

6.2 Significant Achievements, Conclusions and Avenues for Future Research

The research work discussed in this book is the very first attempt to explore and establish WSEM as a superior alternative to conventional processes of manufacturing miniature gears. Successful machining of miniature gears with high quality (DIN-5) and fine surface finish (with 1.05 μm average roughness and 6.34 μm maximum roughness) using WSEM are the two major achievements of the research work discussed. Furthermore, the miniature gears machined at optimum WSEM parameters are defect-free and have very thin recast layer with good microstructure.

More importantly, the quality obtained is without the assistance of any finishing operation, and eliminating the need of subsequent finishing which therefore implies a short process chain with the associated benefits of increased sustainability and makes this process to be recognized as a sustainable gear manufacturing technique.

The results of comparative study and the gear quality obtained prove the superiority and capability of WSEM to overcome the limitations of conventional processes and to manufacture net shaped or near-net shaped gears for high quality miniaturized devices including MEMS, business machines, micro-motors and micro-pumps, scientific and bio-medical instruments etc.

Conclusively, WSEM conducted on an appropriate machine tool coupled with optimized process parameter settings can be used to manufacture miniature gears of excellent quality and good surface integrity aspects at par with those produced by grinding, honing, lapping and other finishing processes with minimum wastage and at comparatively low cost. Reasonable constraint on the discharge energy and wire vibration is the key to machining quality miniature gears by WSEM.

Since, the present work is probably the first attempt to explore in detail and establish WSEM as an alternative process for near-net shape manufacturing of miniature gears; therefore, enough scope remains to conduct the future research in this area. Avenues for future research may include investigating the effects of some other WSEM parameters such as wire tension, wire materials, flushing pressure and discharge current on the quality of miniature gears; exploring other materials such as stainless steel, bronze and aluminium etc. for WSEM of miniature gears; attempting gear tooth modifications i.e. tip correction, root alteration and finishing of gear teeth using WSEM etc.

References

1. Townsend DP (2011) Dudley's gear handbook. Tata McGraw-Hill Publishing Company, New Delhi
2. Gupta K, Jain NK (2014) Comparative study of wire-EDM and hobbing for manufacturing high quality miniature gears. Mater Manuf Processes 29:1470–1476
3. Gimpert D (2007) Gear inspection. Gear Solutions Mag, 36–43
4. Gupta K, Jain NK (2014) On surface integrity of miniature spur gears manufactured by wire electrical discharge machining. Int J Adv Manuf Technol 72(9–12):1735–1745
5. Gupta K, Jain NK (2013) Deviations in geometry of miniature gears fabricated by wire electrical discharge machining. In: Proceedings of international mechanical engineering congress & exposition (IMECE 2013) of ASME, V010T11A047, San Diego, USA, 13–21 Nov 2013
6. Gupta K, Laubscher RF, Davim JP, Jain NK (2016) Recent developments in sustainable manufacturing of gears-a review. J Clean Prod 112(4):3320–3330
7. Gupta K, Laubscher RF, Jain NK (2015) Experimental investigations on manufacturing of high quality miniature gears by wire electric discharge machining. In: Proceedings of international conference on gear production, Munich (Germany), 5–6 October 2015, pp 1471–1482
8. Gupta K, Chaubey SK, Jain NK (2014) Exploring wire-EDM for manufacturing the high quality meso-gears. Proc Mater Sci 5:1755–1760

Appendix A

Chemical composition of gear material (ASTM 858 Brass)

Element	Quantity (%)
Cu	59.30
Zn	38.136
Sn	0.166
Pb	1.254
Mn	0.047
Al	0.296
Fe	0.383
Si	0.009
Ni	0.305
P	0.104

Based on Optical Emission Spectrometry

© Springer Science+Business Media Singapore 2017

K. Gupta and N.K. Jain, *Near-Net Shape Manufacturing of Miniature Spur Gears by Wire Spark Erosion Machining*, Materials Forming, Machining and Tribology, DOI 10.1007/978-981-10-1563-2

Appendix B

CNC program for WSEM of miniature gear using 250 μm diameter wire

```
G71
G9
G27
G40
G47
G50
G90
G75
; Wire Compensation Definitions
D0=0
D1=0.175
;#1.0 Cavity=1 RoughCut
G0 X0 Y0 U0 V0
M21
G42 D0 ;D0=0
G1 X0 Y-0.618747
G42 D1 ;D1=0.175
G3 X-0.217243 Y-0.623566 I0 J-4.9
G3 X-0.366642 Y-0.862402 I1.478333 J-1.090884
G3 X-0.47777 Y-1.121271 I1.627732 J-0.852047
G3 X-0.54249 Y-1.364312 I1.738861 J-0.593178
G3 X-0.57341 Y-1.613915 I1.80358 J-0.350137
G1 X-0.483083 Y-2.229028
G3 X-0.860573 Y-2.307044 I0.483083 J-3.28972
G3 X-1.226498 Y-2.428225 I0.860573 J-3.211703
G1 X-1.455829 Y-1.850359
G3 X-1.607408 Y-1.649656 I-1.538457 J-1.004315
G3 X-1.784977 Y-1.471536 I-1.386878 J-1.205017
G3 X-2.010652 Y-1.302913 I-1.209308 J-1.383138
G3 X-2.259453 Y-1.170774 I-0.983634 J-1.551761
G3 X-2.45 Y-1.275223 I2.259453 J-4.347973
```


© Springer Science+Business Media Singapore 2017
K. Gupta and N.K. Jain, *Near-Net Shape Manufacturing of Miniature Spur Gears by Wire Spark Erosion Machining*, Materials Forming, Machining and Tribology, DOI 10.1007/978-981-10-1563-2

```
G3 X-2.635729 Y-1.388017 I2.45 J-4.243524
G3 X-2.645694 Y-1.669555 I1.825716 J-0.205566
G3 X-2.6125 Y-1.949306 I1.835681 J0.075971
G3 X-2.547027 Y-2.192146 I1.802487 J0.355723
G3 X-2.449003 Y-2.423768 I1.737015 J0.598563
G1 X-2.063222 Y-2.911308
G3 X-2.35113 Y-3.167617 I2.063222 J-2.60744
G3 X-2.60744 Y-3.455526 I2.35113 J-2.35113
G1 X-3.094979 Y-3.069744
G3 X-3.326602 Y-2.97172 I-0.830185 J-1.63899
G3 X-3.569441 Y-2.906248 I-0.598563 J-1.737015
G3 X-3.849193 Y-2.873053 I-0.355723 J-1.802487
G3 X-4.13073 Y-2.883018 I-0.075971 J-1.835681
G3 X-4.243524 Y-3.068747 I4.13073 J-2.635729
G3 X-4.347973 Y-3.259295 I4.243524 J-2.45
G3 X-4.215835 Y-3.508096 I1.6839 J0.734833
G3 X-4.047212 Y-3.73377 I1.551761 J0.983634
G3 X-3.869091 Y-3.91134 I1.383138 J1.209308
G3 X-3.668389 Y-4.062919 I1.205017 J1.386878
G1 X-3.090522 Y-4.292249
G3 X-3.211703 Y-4.658174 I3.090522 J-1.226498
G3 X-3.28972 Y-5.035665 I3.211703 J-0.860573
G1 X-3.904832 Y-4.945338
G3 X-4.154435 Y-4.976258 I0.100534 J-1.8345
G3 X-4.397477 Y-5.040977 I0.350137 J-1.80358
G3 X-4.656346 Y-5.152106 I0.593178 J-1.738861
G3 X-4.895182 Y-5.301504 I0.852047 J-1.627732
G3 X-4.9 Y-5.518747 I4.895182 J-0.217243
G3 X-4.895182 Y-5.735991 I4.9 J0
G3 X-4.656346 Y-5.885389 I1.090884 J1.478333
G3 X-4.397477 Y-5.996518 I0.852047 J1.627732
G3 X-4.154435 Y-6.061237 I0.593178 J1.738861
G3 X-3.904832 Y-6.092157 I0.350137 J1.80358
G1 X-3.28972 Y-6.00183
G3 X-3.211703 Y-6.379321 I3.28972 J0.483083
G3 X-3.090522 Y-6.745245 I3.211703 J0.860573
G1 X-3.668389 Y-6.974576
G3 X-3.869091 Y-7.126155 I1.004315 J-1.538457
G3 X-4.047212 Y-7.303725 I1.205017 J-1.386878
G3 X-4.215835 Y-7.529399 I1.383138 J-1.209308
G3 X-4.347973 Y-7.7782 I1.551761 J-0.983634
G3 X-4.243524 Y-7.968747 I4.347973 J2.259453
G3 X-4.13073 Y-8.154476 I4.243524 J2.45
G3 X-3.849193 Y-8.164441 I0.205566 J1.825716
G3 X-3.569441 Y-8.131247 I-0.075971 J1.835681
```

```
G3 X-3.326602 Y-8.065775 I-0.355723 J1.802487
G3 X-3.094979 Y-7.967751 I-0.598563 J1.737015
G1 X-2.60744 Y-7.581969
G3 X-2.35113 Y-7.869877 I2.60744 J2.063222
G3 X-2.063222 Y-8.126187 I2.35113 J2.35113
G1 X-2.449003 Y-8.613726
G3 X-2.547027 Y-8.845349 I1.63899 J-0.830185
G3 X-2.6125 Y-9.088189 I1.737015 J-0.598563
G3 X-2.645694 Y-9.36794 I1.802487 J-0.355723
G3 X-2.635729 Y-9.649478 I1.835681 J-0.075971
G3 X-2.45 Y-9.762272 I2.635729 J4.13073
G3 X-2.259453 Y-9.866721 I2.45 J4.243524
G3 X-2.010652 Y-9.734582 I-0.734833 J1.6839
G3 X-1.784977 Y-9.565959 I-0.983634 J1.551761
G3 X-1.607408 Y-9.387839 I-1.209308 J1.383138
G3 X-1.455829 Y-9.187136 I-1.386878 J1.205017
G1 X-1.226498 Y-8.60927
G3 X-0.860573 Y-8.730451 I1.226498 J3.090522
G3 X-0.483083 Y-8.808467 I0.860573 J3.211703
G1 X-0.57341 Y-9.42358
G3 X-0.54249 Y-9.673183 I1.8345 J0.100534
G3 X-0.47777 Y-9.916224 I1.80358 J0.350137
G3 X-0.366642 Y-10.175093 I1.738861 J0.593178
G3 X-0.217243 Y-10.413929 I1.627732 J0.852047
G3 X0 Y-10.418747 I0.217243 J4.895182
G3 X0.217243 Y-10.413929 I0 J4.9
G3 X0.366642 Y-10.175093 I-1.478333 J1.090884
G3 X0.47777 Y-9.916224 I-1.627732 J0.852047
G3 X0.54249 Y-9.673183 I-1.738861 J0.593178
G3 X0.57341 Y-9.42358 I-1.80358 J0.350137
G1 X0.483083 Y-8.808467
G3 X0.860573 Y-8.730451 I-0.483083 J3.28972
G3 X1.226498 Y-8.60927 I-0.860573 J3.211703
G1 X1.455829 Y-9.187136
G3 X1.607408 Y-9.387839 I1.538457 J1.004315
G3 X1.784977 Y-9.565959 I1.386878 J1.205017
G3 X2.010652 Y-9.734582 I1.209308 J1.383138
G3 X2.259453 Y-9.866721 I0.983634 J1.551761
G3 X2.45 Y-9.762272 I-2.259453 J4.347973
G3 X2.635729 Y-9.649478 I-2.45 J4.243524
G3 X2.645694 Y-9.36794 I-1.825716 J0.205566
G3 X2.6125 Y-9.088189 I-1.835681 J-0.075971
G3 X2.547027 Y-8.845349 I-1.802487 J-0.355723
G3 X2.449003 Y-8.613726 I-1.737015 J-0.598563
G1 X2.063222 Y-8.126187
```

```
G3 X2.35113 Y-7.869877 I-2.063222 J2.60744
G3 X2.60744 Y-7.581969 I-2.35113 J2.35113
G1 X3.094979 Y-7.967751
G3 X3.326602 Y-8.065775 I0.830185 J1.63899
G3 X3.569441 Y-8.131247 I0.598563 J1.737015
G3 X3.849193 Y-8.164441 I0.355723 J1.802487
G3 X4.13073 Y-8.154476 I0.075971 J1.835681
G3 X4.243524 Y-7.968747 I-4.13073 J2.635729
G3 X4.347973 Y-7.7782 I-4.243524 J2.45
G3 X4.215835 Y-7.529399 I-1.6839 J-0.734833
G3 X4.047212 Y-7.303725 I-1.551761 J-0.983634
G3 X3.869091 Y-7.126155 I-1.383138 J-1.209308
G3 X3.668389 Y-6.974576 I-1.205017 J-1.386878
G1 X3.090522 Y-6.745245
G3 X3.211703 Y-6.379321 I-3.090522 J1.226498
G3 X3.28972 Y-6.00183 I-3.211703 J0.860573
G1 X3.904832 Y-6.092157
G3 X4.154435 Y-6.061237 I-0.100534 J1.8345
G3 X4.397477 Y-5.996518 I-0.350137 J1.80358
G3 X4.656346 Y-5.885389 I-0.593178 J1.738861
G3 X4.895182 Y-5.735991 I-0.852047 J1.627732
G3 X4.9 Y-5.518747 I-4.895182 J0.217243
G3 X4.895182 Y-5.301504 I-4.9 J0
G3 X4.656346 Y-5.152106 I-1.090884 J-1.478333
G3 X4.397477 Y-5.040977 I-0.852047 J-1.627732
G3 X4.154435 Y-4.976258 I-0.593178 J-1.738861
G3 X3.904832 Y-4.945338 I-0.350137 J-1.80358
G1 X3.28972 Y-5.035665
G3 X3.211703 Y-4.658174 I-3.28972 J-0.483083
G3 X3.090522 Y-4.292249 I-3.211703 J-0.860573
G1 X3.668389 Y-4.062919
G3 X3.869091 Y-3.91134 I-1.004315 J1.538457
G3 X4.047212 Y-3.73377 I-1.205017 J1.386878
G3 X4.215835 Y-3.508096 I-1.383138 J1.209308
G3 X4.347973 Y-3.259295 I-1.551761 J0.983634
G3 X4.243524 Y-3.068747 I-4.347973 J-2.259453
G3 X4.13073 Y-2.883018 I-4.243524 J-2.45
G3 X3.849193 Y-2.873053 I-0.205566 J-1.825716
G3 X3.569441 Y-2.906248 I0.075971 J-1.835681
G3 X3.326602 Y-2.97172 I0.355723 J-1.802487
G3 X3.094979 Y-3.069744 I0.598563 J-1.737015
G1 X2.60744 Y-3.455526
G3 X2.35113 Y-3.167617 I-2.60744 J-2.063222
G3 X2.063222 Y-2.911308 I-2.35113 J-2.35113
G1 X2.449003 Y-2.423768
```

```
G3 X2.547027 Y-2.192146 I-1.63899 J0.830185
G3 X2.6125 Y-1.949306 I-1.737015 J0.598563
G3 X2.645694 Y-1.669555 I-1.802487 J0.355723
G3 X2.635729 Y-1.388017 I-1.835681 J0.075971
G3 X2.45 Y-1.275223 I-2.635729 J-4.13073
G3 X2.259453 Y-1.170774 I-2.45 J-4.243524
G3 X2.010652 Y-1.302913 I0.734833 J-1.6839
G3 X1.784977 Y-1.471536 I0.983634 J-1.551761
G3 X1.607408 Y-1.649656 I1.209308 J-1.383138
G3 X1.455829 Y-1.850359 I1.386878 J-1.205017
G1 X1.226498 Y-2.428225
G3 X0.860573 Y-2.307044 I-1.226498 J-3.090522
G3 X0.483083 Y-2.229028 I-0.860573 J-3.211703
G1 X0.57341 Y-1.613915
G3 X0.54249 Y-1.364312 I-1.8345 J-0.100534
G3 X0.47777 Y-1.121271 I-1.80358 J-0.350137
G3 X0.366642 Y-0.862402 I-1.738861 J-0.593178
G3 X0.217243 Y-0.623566 I-1.627732 J-0.852047
G3 X0 Y-0.618747 I-0.217243 J-4.895182
M0
G42 D0 ;D0=0
G1 X0 Y0
G40
M0
```

Index

© Springer Science+Business Media Singapore 2017
K. Gupta and N.K. Jain, *Near-Net Shape Manufacturing of Miniature Spur Gears by Wire Spark Erosion Machining*, Materials Forming, Machining and Tribology, DOI 10.1007/978-981-10-1563-2

MIX
Papier aus verantwortungsvollen Quellen
Paper from responsible sources
FSC® C105338

If you have any concerns about our products,
you can contact us on
ProductSafety@springernature.com

In case Publisher is established outside the EU,
the EU authorized representative is:
Springer Nature Customer Service Center GmbH
Europaplatz 3, 69115 Heidelberg, Germany

Printed by Libri Plureos GmbH
in Hamburg, Germany